INTRODUCTION TO SAN ISABEL NATIONAL FOREST

WELCOME

Welcome to the San Isabel National Forest, one of eleven National Forests in Colorado. The Forest includes over one million acres of beautiful scenery with snow-capped mountains, wildflowers, autumn colors, mountain lakes, and clear blue skies to enjoy.

Prior to the establishment of the Forest, the presence of Indians, Spanish Land Grants, homesteading and the discovery of gold were important in shaping the land. The lands originally set aside as a Forest Reserve in 1902 were renamed San Isabel National Forest in March, 1907. From 1907 until 1945 the Forest grew steadily in size, as several other Forests and additional lands were integrated into the San Isabel National Forest. Today the Forest is administered by three District Offices and the Supervisor's Office.

The Forest is bounded on the west and north by the Continental Divide, and on the east by the Pike National Forest. The Wet Mountains, Collegiate Peaks, Sawatch Range, Spanish Peaks and the Sangre De Cristos provide a variety of scenery. Elevations range from a low of 5,860 feet to 14,433. The highest point is Mount Elbert, Colorado's highest peak. The high elevations account for the comfortable summer temperatures and year round snow on the higher peaks.

Another attraction is the Arkansas Headwaters Recreation Area. The Recreation Area which stretches for 148 miles along the Arkansas, begins in Leadville and provides numerous opportunities for fishing, rafting, picnicking and camping. The Recreation Area is managed by the Bureau of Land Management and the Colorado Division of Parks and Outdoor Recreation, (719-539-7289).

Black bear, mule deer, elk, bighorn sheep, mountain goats, turkey, mountain lions are among the animals and birds that make their homes within the Forest.

Today, almost 800 miles of trails, ski areas, nineteen peaks over 14,000 feet, scenic byways, numerous roads and highways, campgrounds and picnic areas provide challenges and opportunities for everyone.

All this is yours to enjoy, but please, do so in ways that will allow others to enjoy it today and tomorrow!

CLIMATE

Weather on the San Isabel National Forest is as varied as the topography. Extreme variations can occur; yearly, daily, or even hourly. Weather is influenced by the terrain and high elevations of the Rocky Mountains.

Elevations ranging from 6,000 to over 14,000 feet reduce oxygen in the air, thus making it harder to breathe. Sunburns occur easily at this elevation due to the lack of air filter ultraviolet radiation. Sunscreen is highly recommended. Elevation also affects temperature, cooling about 3 degrees fahrenheit for each 1,000 feet gained in elevation. Snowfall is possible anytime of the year.

Come prepared when visiting the San Isabel National Forest. Thunderstorms and lightening are frequent during the summer, especially on the ridges. Avoid these locations during stormy weather. Freezing temperatures at night are common throughout the summer. Layers of clothing are best when preparing for a trip or vacation.

Hold on to your hat while visiting the San Isabel National Forest, for it can be quite windy. Wind is often strong, coming primarily from the west.

CONTINENTAL DIVIDE AND COLORADO TRAILS

Two long-distance hiking trails of national importance pass through the San Isabel National Forest: the Colorado Trail and the Continental Divide National Scenic Trail. The Colorado Trail is a 469-mile trail between Denver and Durango. The 3,100-mile Continental Divide Trail is proposed between the Mexican and Canadian borders. Numerous side trails connect these two trails with campgrounds, trailheads and local communities along the way.

The Colorado Trail is open to hikers, horseback riders, and cross-country skiers. Mountain bikes are currently permitted on the non-Wilderness segments of the trail. The trail passes through seven National Forests, six Wildernesses, five major river systems and eight mountain ranges. The trail is even more impressive because it was created through a massive volunteer effort involving thousands of trail-building enthusiasts.

The Colorado Trail is jointly administered by the Colorado Trail Foundation, and the U.S. Forest Service.

The 3,100-mile Continental Divide National Scenic Trail, when completed will provide spectacular backcountry travel the length of the Rocky Mountains from Mexico to Canada. It is the most rugged of the long-distance trails. The trail was designated in 1978 and is designed to be an educational as well as a hiking experience. The trail traverses a variety of terrain, including high desert, forests, geologic formations and mountian meadows. Along the way, travelers can catch glimpses of a variety of historical, cultural and scenic landscapes, as well as abundant wildlife. Hiking opportunities range from short leisurely hikes to challenging alpine and desert segments. On the San Isabel National Forest, one of 25 National Forests traversed by the Continental Divide Trail, 100 miles of the trail are between Tennessee Pass on the north and Windy Peak on the south.

Both the Colorado Trail and Continental Divide National Scenic Trail provide recreation opportunities and preserve a part of America for future generations.

U.S. Forest Service

FOREST SUPERVISOR
Pike & San Isabel National Forests
1920 Valley Drive
Pueblo, CO 81008
(719) 545-8737

TABLE of CONTENTS

	Page
SECTION 1 - TRAILS	6
SECTION 2 - BACKPACKING IS FREEDOM	147
SECTION 3 - CAMPING	153
TURQUOISE LAKE	142
TWIN LAKES	144
APPENDIX	160
METHOD FOR RATING TRAIL DIFFICULTY	159

How To Use This Guide
Trails described in the text are shown on the map page opposite the text. The map number shown on this page refers to the map number shown on the index map. The index map located in the front of the guide shows map numbers relative to the major roads and towns within the National Forest.

ACKNOWLEDGMENTS
WE WOULD LIKE TO THANK THE U. S. FOREST SERVICE FOR THEIR COOPERATION AND ASSISTANCE FOR MAKING THIS GUIDE POSSIBLE.

Most information contained in this document was compiled from Forest Service Recreation Opportunity Guides.

U.S. Geologic Survey
Colorado Division of State Parks and Outdoor Recreation

Cover Photo: **Tom Myers**
Cover Inset Photos: **PhotoDisc**
Cover Design & Production: **Ken Grasman**
Map Graphics: **Michael Moore**
Production Artist: **Linda Bollinger**

Editor & Publisher
Jack O. Olofson

Staff
Kristin Alexander
Ryan Alexander
Cynthia Alexander
Linnea Roberts
Dody Olofson

© Copyright 1995 Outdoor Books & Maps, Inc. - Denver Colorado. This publication can not be reproduced in whole or part without written permission from Outdoor Books & Maps, Inc.

P.O. Box 417 Denver, Colorado 80201
Phone (303) 629-6111 • Fax (303) 628-9378

EXPLANATION

Map Scale: 1" = approximately 2 miles

P Parking Area	Forest Service Facility	(285) U.S. Highway	National Forest Area
Picnic Area	Fishing Area	(126) State Highway	Water
Trail Head	Snowmobile Trail	(258) County Highway	Trail
Downhill Ski Area	Ice Skating Area	(1606) Trail Number	River or Stream
Boat Launch	Cross Country Ski Area	(1105) Forest Service Road	Primary Road - Paved
Bicycle Trail	Hunting	▲ Mountain	Improved Road - Unpaved
4WD Road	❶△ Campground	Colorado Trail	Unimproved Road - 4WD
Motorcycle Trail	Towns & Locales	Continental Divide Trail	Forest / Wilderness Boundary

Off-Highway Vehicles (OHV's)

On National Forest land all OHV's must be equipped with operating brakes, a muffler and a Forest Service-approved spark arrester. OHVs must also have and use a headlight and taillight between sunset and sunrise. OHVs should also be operated in a safe, quiet manner without disturbing the land, vegetation or wildlife.

Mountain Bikes:

Roads and most trails, with the exception of those in Wildernesses, are open for mountain bike use. Some trails are unsuitable due to steep grades, rocky conditions or stream crossings.

In order to have a safe and enjoyable ride, please keep the following suggestions in mind:
1. Trail challenges vary, so know the terrain and your skill level.
2. Be courteous to others on the trail.
3. When on the road, obey traffic regulations.
4. Stay on the road or trail, avoid shortcuts, switchbacks and damage to vegetation.
5. Wear a helmet.
6. Keep your bike in good repair and carry proper tools and spare parts.

Campgrounds:

Reservations: Although many Forest Service campgrounds are available on a first-come, first-served basis, selected sites and group areas are on a reservation system through a private corporation. The local District Ranger's office can tell you which campgrounds are on the reservation system, but they cannot make reservations for you. For information and reservations call Mistix 1-800-283-CAMP

Fees: Most campgrounds are fee areas. They will be clearly posted as such. Please follow posted instructions carefully and promptly. Closures: In some high-use areas, camping is restricted to developed sites in order to avoid erosion, sanitation, and fire safety problems. Please be alert for posted regulations in these high-use areas.

Dispersed Camping: When camping or picnicking outside developed sites, please use minimum-impact techniques. Leave a campsite as nice or nicer than you found it.

Wilderness:

In order to preserve wilderness areas for others to enjoy please follow these "no trace" camping guidelines.
1. Travel in small groups, avoiding heavy use areas.
2. Pack out metal, glass and other unburnables.
3. Don't bathe, wash dishes or camp within 100 feet of streams.
4. Don't tie horses or pack animals to trees.
5. Set up camp where you'll create the least disturbance.
6. Backpack stoves are recommended.
7. Human waste should be buried, and be at least 100 feet from streams.

General:

All Forest Service roads and trails that are open to motorized vehicles are signed on the ground with white arrows. This signing should be used to confirm the motorized travel status of all Forest routes.

Except as noted, the San Isabel National Forest is closed to cross-country motorized use in order to prevent resource damage, such as that caused by the unplanned development of new roads, and to reduce the disturbance of wildlife.

NUMERICAL TRAILS LISTING

Trails are in numerical order in document.

	Page
Trail No. **1300** — *Indian*	6
Trail No. **1301** — *Chaparral*	8
Trail No. **1303** — *Cascade Creek*	1
Trail No. **1304** — *Wahatoya*	12
Trail No. **1306** — *Ute*	14
Trail No. **1308** — *Lilly Lake*	16
Trail No. **1309** — *North Fork*	18
Trail No. **1310** — *Barlett*	20
Trail No. **1312** — *Baker*	22
Trail No. **1314** — *Cisneros*	24
Trail No. **1316** — *Greenhorn*	26
Trail No. **1318** — *Snowslide*	28
Trail No. **1321** — *South Creek*	30
Trail No. **1322** — *Second Mace*	32
Trail No. **1323** — *Silver Circle*	34
Trail No. **1325** — *North Creek*	36
Trail No. **1326** — *Saint Charles*	38
Trail No. **1327** — *Rudloph Mountain*	40
Trail No. **1331** — *Lewis Creek*	42
Trail No. **1332** — *Mineral - Stevens*	44
Trail No. **1333** — *Tanner*	46
Trail No. **1336** — **Rainbow Trail**	
Segment A-B — *Grape Creek - Horn Creek*	48
Segment B-C — *Horn Cr. - Middle Taylor Cr.*	50
Segment C-D — *Middle Taylor Creek - North Brush Creek*	52
Segment D-E — *North Brush Creek - Big Cottonwood Creek*	54
Segment E-F — *Big Cottonwood Cr. - Stout Cr.*	56
Segment F-G — *Stout Creek - Bear Creek*	58
Segment G-H — *Bear Creek - Hwy 285*	60
Segment H-I — *Hwy 285 - Colo. Trail #1776*	62
Trail No. **1339** — *South Colony*	64
Trail No. **1340** — *North Colony*	66
Trail No. **1341** — *Macey*	68
Trail No. **1342** — *Horn Creek*	70
Trail No. **1343** — *Dry Creek*	72

	Page
Trail No. **1344** — *Cottonwood*	74
Trail No. **1345** — *Comanche-Venable*	76
Trail No. **1346** — *Goodwin Lake*	78
Trail No. **1349** — *Lake of The Clouds*	80
Trail No. **1355** — *South Brush Creek*	82
Trail No. **1356** — *North Brush Creek*	84
Trail No. **1374** — *Pine Creek*	86
Trail No. **1378** — *Missouri Gulch*	88
Trail No. **1384** — *Squirrel Creek*	90
Trail No. **1387** — *Dome Rock*	92
Trail No. **1390** — *West Spanish Peak*	94
Trail No. **1427** — *Wagon Road Loop*	96
Trail No. **1429** — *Browns Creek*	98
Trail No. **1436** — *Poplar Gulch*	100
Trail No. **1439** — *Tunnel Lake*	102
Trail No. **1444** — *Ptarmigan Lake*	104
Trail No. **1449** — *Dodgeton*	106
Trail No. **1469** — *Missouri Mountain*	108
Trail No. **1474** — *La Plata Peak*	110
Trail No. **1481** — *South Mt. Elbert*	112
Trail No. **1484** — *North Mt. Elbert*	114
Trail No. **1494** — *Mount Columbia*	116
Trail No. **1498** — *Mount Huron*	118
Trail No. **1501** — *Mount Harvard*	120
Trail No. **1776** — **Colorado Trail**	
Segment A-B — *Tennessee Pass to Turquoise Lake*	122
Segment B-C — *Turquoise Lake to Halfmoon Cr.*	124
Segment C-D — *Halfmoon Creek to Twin Lakes*	126
Segment D-E — *Twin Lakes to Clear Creek*	128
Segment E-F — *Clear Creek to North Cottonwood Creek*	130
Segment F-G — *No Cottonwood Creek to South Cottonwood Crk*	132
Segment G-H — *So Cottonwood Cr. to Chalk Cr.*	134
Segment H-I — *Marshall Pass to Sargents Mesa*	136
Segment I-J — *Chalk Creek to US Hwy 50*	138
Segment J-K — *US Hwy 50 to Marshall Pass*	140

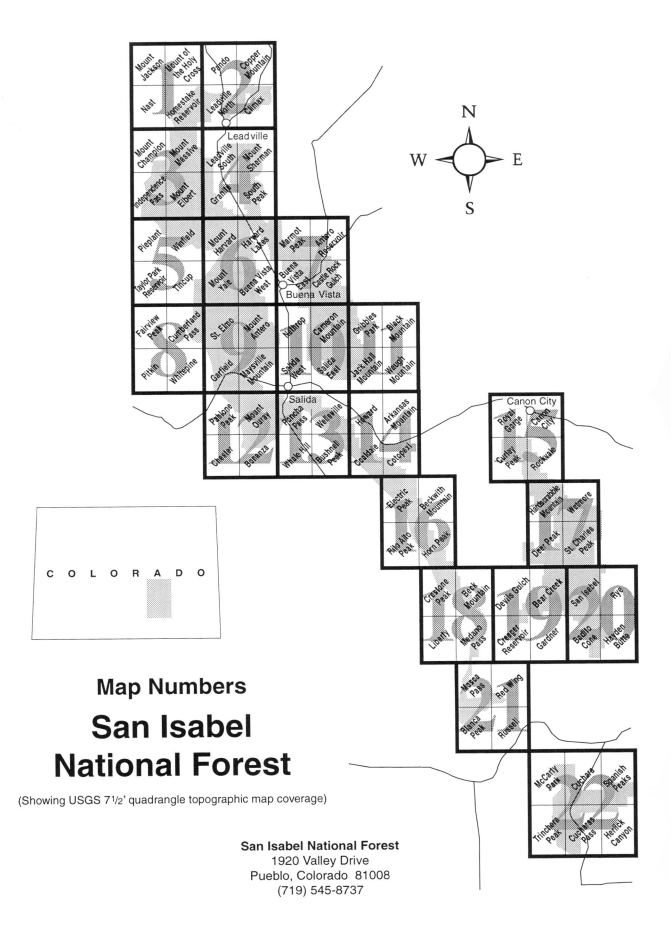

Map Numbers
San Isabel National Forest

(Showing USGS 7½' quadrangle topographic map coverage)

San Isabel National Forest
1920 Valley Drive
Pueblo, Colorado 81008
(719) 545-8737

ROCKY MOUNTAIN REGION
MAP NUMBER(S): 22

NATIONAL FOREST: SAN ISABEL
RANGER DISTRICT: SAN CARLOS

USGS: MCCARTY, TRINCHERA PEAK QUADS

DIFFICULTY: EASY

DISTANCE: 5.0 MILES

TRAIL 1300
INDIAN CREEK

UPDATED:

TRAIL BEGINNING ELEVATION:
10,540 ft. Bear Lake Campground

TRAIL ENDING ELEVATION:
9,480 ft. Intersection of Trail #1302

ACCESS:
18 miles south of La Veta on Highway 12 then go 3 miles on FDR 422 to Bear Lake Campground on FDR #413.

ATTRACTIONS:
Scenery, facilities at Bear Lake Campground. Fishing at Blue and Bear Lakes.

USE: Heavy

RECOMMENDED SEASON:

SPRING	SUMMER	FALL	WINTER

NARRATIVE:

TRAIL PROFILE (ONE-WAY):

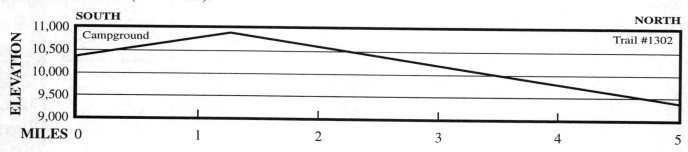

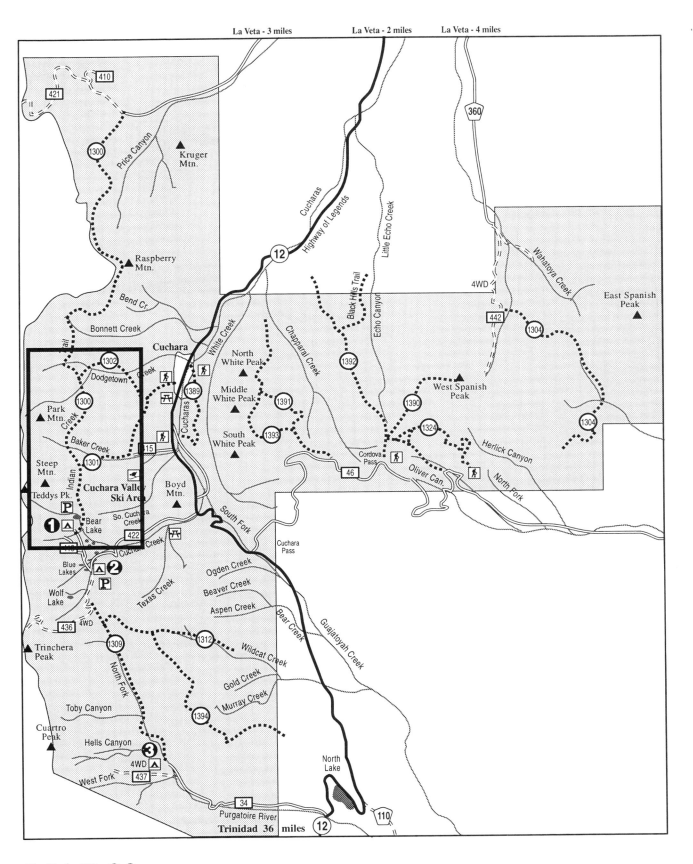

MAP 22
TRAIL #1300

© 1995 — *Outdoor Books & Maps, Inc.* • *Denver, Colorado* • (303) 629-6111

ROCKY MOUNTAIN REGION

MAP NUMBER(S): 22

NATIONAL FOREST: SAN ISABEL
RANGER DISTRICT: SAN CARLOS

USGS: CUCHARA, CUCHARA PASS, TRINCHERA PEAK QUADS

UPDATED:

TRAIL BEGINNING ELEVATION:
8,600 ft. Cuchara

TRAIL ENDING ELEVATION:
10,300 ft. Intersection of Trail #1300

ACCESS:
10 miles south of La Veta on Highway 12 to FSR 415 to trailhead.

ATTRACTIONS:
Scenery, no facilities.

DIFFICULTY: EASY

DISTANCE: 4.0 MILES

TRAIL 1301 CHAPPARRAL

USE: Moderate

RECOMMENDED SEASON: SPRING | **SUMMER | FALL** | WINTER

NARRATIVE:

TRAIL PROFILE (ONE-WAY):

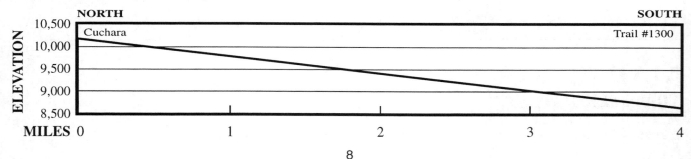

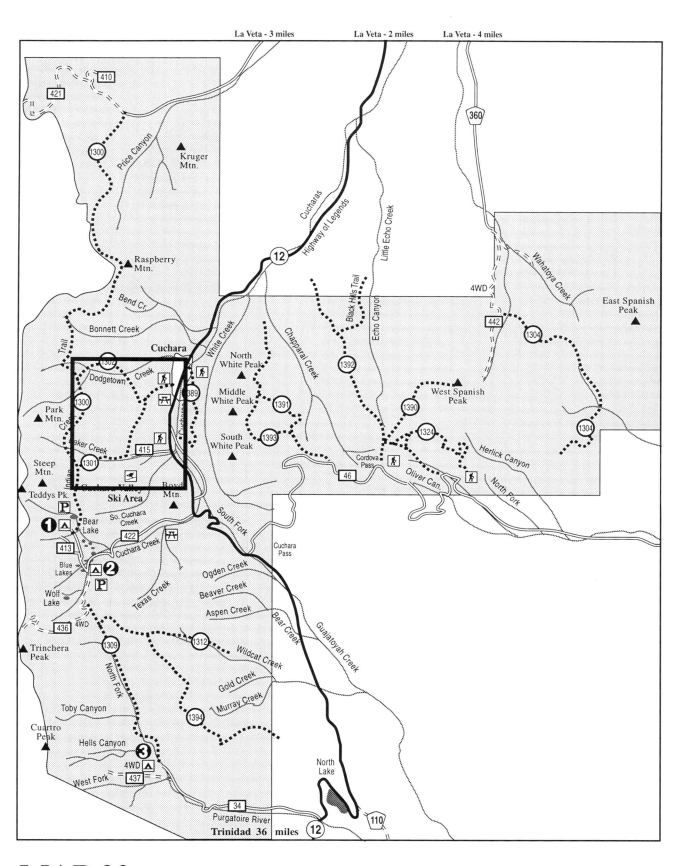

MAP 22
TRAIL #1301

© 1995 — *Outdoor Books & Maps, Inc.* • *Denver, Colorado* • (303) 629-6111

ROCKY MOUNTAIN REGION

MAP NUMBER(S): 21

NATIONAL FOREST: SAN ISABEL
RANGER DISTRICT: SAN CARLOS

USGS: MOSCA PASS, RED WING QUADS

DIFFICULTY: DIFFICULT

DISTANCE: 4.0 MILES

TRAIL 1303
CASCADE CREEK

UPDATED:

TRAIL BEGINNING ELEVATION:
9,800 ft. Intersection of Trail #1307

TRAIL ENDING ELEVATION:
10,000 ft. Private property

ACCESS:
7 miles southwest of Gardner on County. Rd. #550 to County Rd. #580. Forward 9 miles through Hueferno State Wildlife Area to 4WD FS Rd. #580. Continue 5 miles to Trail #1307. Proceed 3.5 miles on #1307 to intersection of Trail #1303.

ATTRACTIONS:
Scenery, no facilities.

USE: Moderate

RECOMMENDED SEASON:

| SPRING | SUMMER | FALL | WINTER |

NARRATIVE:

TRAIL PROFILE (ONE-WAY):

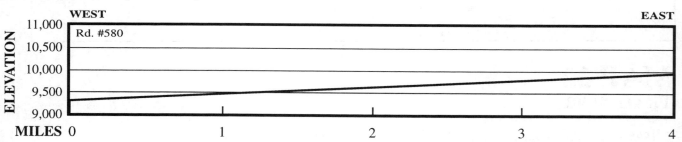

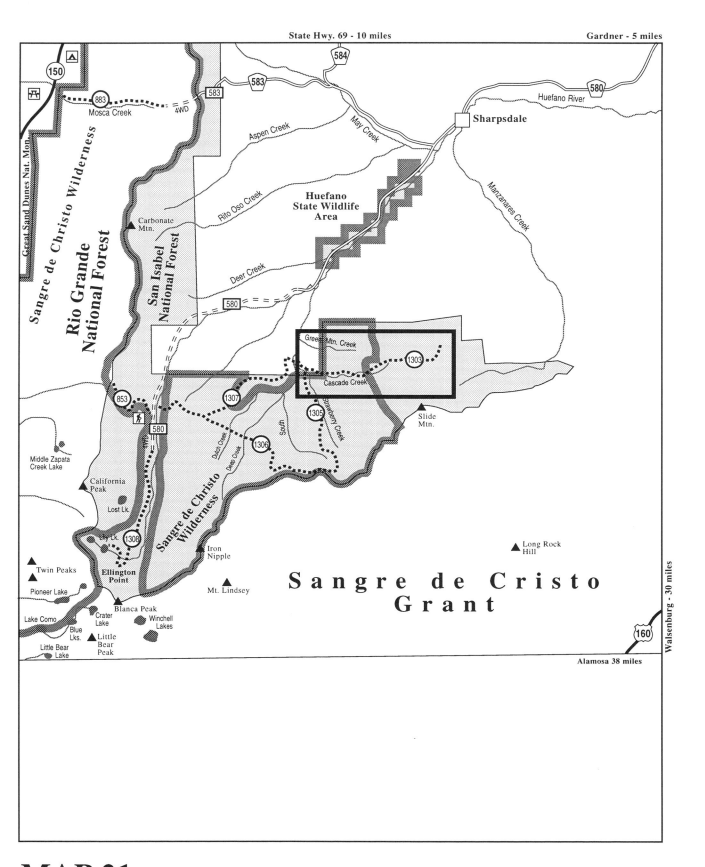

MAP 21
TRAIL #1303

© 1995 — *Outdoor Books & Maps, Inc.* • *Denver, Colorado* • (303) 629-6111

ROCKY MOUNTAIN REGION

MAP NUMBER(S): 22

NATIONAL FOREST: SAN ISABEL
RANGER DISTRICT: SAN CARLOS

USGS: SPANISH PEAKS, HERLICK CANYON QUADS

UPDATED:

TRAIL BEGINNING ELEVATION:
8,400 ft. FS Rd. #442

TRAIL ENDING ELEVATION:
9,000 ft. Private property

ACCESS:
From Le Veta take County Road #361 for 1 mile to County Road #362 1/4 mile to County Road #360. Proceed 5 miles to Wahatoya Camp. Take the 4WD Road #442 1.5 miles to Trail #1304.

ATTRACTIONS:
Scenery, no facilities.

DIFFICULTY: MODERATE

DISTANCE: 7.0 MILES

TRAIL 1304
WAHATOYA

USE: Moderate

RECOMMENDED SEASON:

SPRING	SUMMER	FALL	WINTER
	■	■	

NARRATIVE:

TRAIL PROFILE (ONE-WAY):

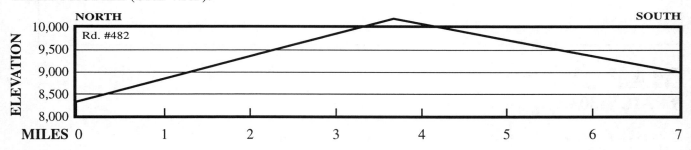

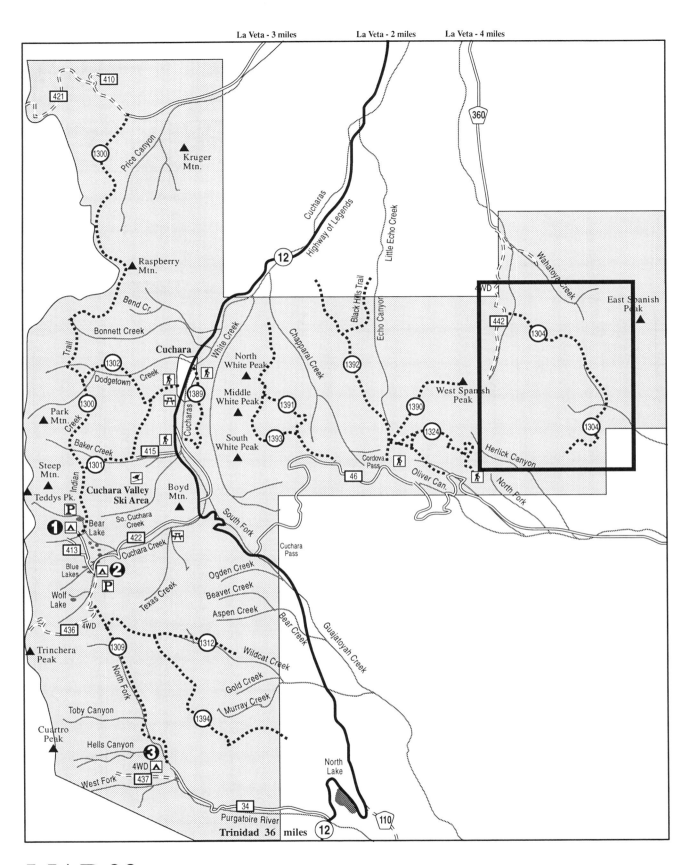

MAP 22
TRAIL #1304

© *1995 — Outdoor Books & Maps, Inc.* • *Denver, Colorado* • **(303) 629-6111**

ROCKY MOUNTAIN REGION

MAP NUMBER(S): 21

NATIONAL FOREST: SAN ISABEL
RANGER DISTRICT: SAN CARLOS

USGS: MOSCA PASS, BLANCA PEAK QUADS

DIFFICULTY: DIFFICULT

DISTANCE: 3.0 MILES

TRAIL 1306
UTE

UPDATED:

TRAIL BEGINNING ELEVATION:
10,940 ft. Trail #1307

TRAIL ENDING ELEVATION:
11,400 ft. Intersection of Trail #1305

ACCESS:
7 miles southwest of Gardner on County Road #550. Proceed 9 miles through Huefano State Wildlife Area to 4WD FS Road #580. Continue 5 miles to Trail #1307. Proceed 2 miles to Trail #1306.

ATTRACTIONS:
Scenery, fishing, no facilities. Sangre de Cristo Wilderness Area.

USE: Moderate

RECOMMENDED SEASON:
SPRING | SUMMER | FALL | WINTER

NARRATIVE:

TRAIL PROFILE (ONE-WAY):

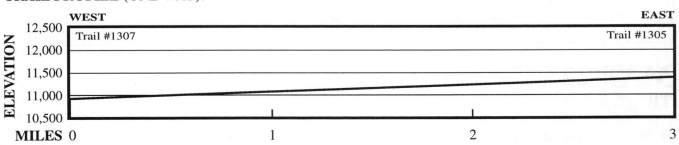

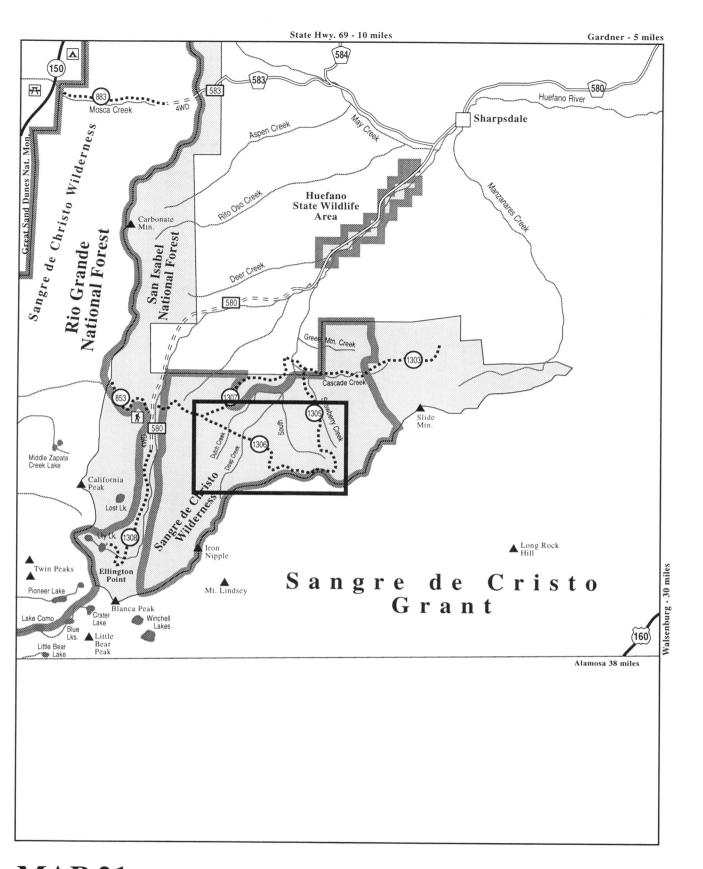

MAP 21
TRAIL #1306

© 1995 — *Outdoor Books & Maps, Inc.* • *Denver, Colorado* • (303) 629-6111

ROCKY MOUNTAIN REGION

MAP NUMBER(S): 21

NATIONAL FOREST: SAN ISABEL
RANGER DISTRICT: SAN CARLOS

USGS: MOSCA PASS, BLANCA PEAK QUADS

DIFFICULTY: DIFFICULT

DISTANCE: 4.0 MILES

TRAIL 1308
LILY LAKE

UPDATED:

TRAIL BEGINNING ELEVATION:
10,400 ft. End of 4WD Road #580

TRAIL ENDING ELEVATION:
12,500 ft. Lily Lake

ACCESS:
7 miles southwest of Gardner on County Road #550 to County Road #580. Proceed 9 miles through Huefano State Wildlife Area to 4WD FS Road #580. Continue 7 miles to end of road and beginning of Trail #1308.

ATTRACTIONS:
Scenery, fishing, no facilities.

USE: Moderate

RECOMMENDED SEASON:

| SPRING | SUMMER | FALL | WINTER |

NARRATIVE:

TRAIL PROFILE (ONE-WAY):

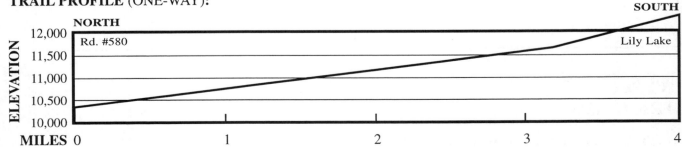

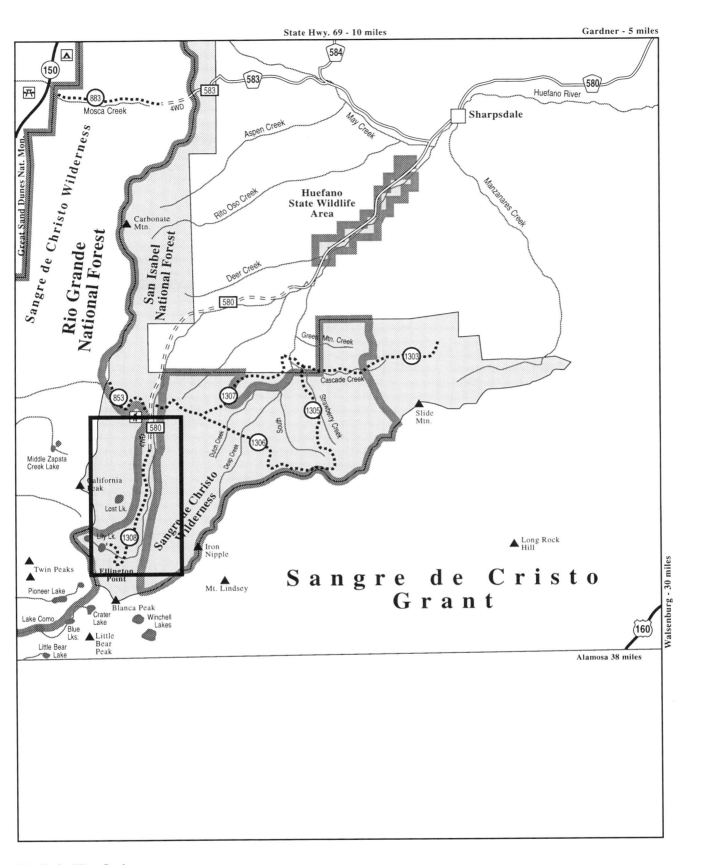

MAP 21
TRAIL #1308

© 1995 — Outdoor Books & Maps, Inc. • Denver, Colorado • (303) 629-6111

ROCKY MOUNTAIN REGION
MAP NUMBER(S): 22

NATIONAL FOREST: SAN ISABEL
RANGER DISTRICT: SAN CARLOS

USGS: CUCHARAS PASS, TRINCHERA PEAK QUADS

DIFFICULTY: EASY

DISTANCE: 4.5 MILES

TRAIL
1309
NORTH FORK

UPDATED:

TRAIL BEGINNING ELEVATION:
9,727 ft. Purgatory Campground

TRAIL ENDING ELEVATION:
11,300 ft. FS Rd. #436

ACCESS:
18 miles from La Veta via Hwy 12, to FS Rd. #422. Continue to intersection of Blue Lake Road #413 and Trinchera Peak Road #436, south to trailhead.

ATTRACTIONS:
Mountain Scenery, parking.

USE: Moderate

RECOMMENDED SEASON:

| SPRING | SUMMER | FALL | WINTER |

NARRATIVE:

TRAIL PROFILE (ONE-WAY):

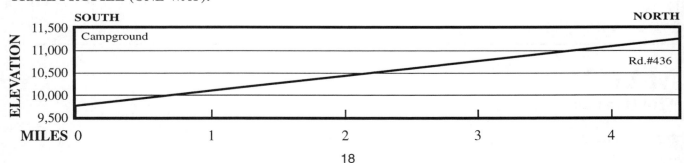

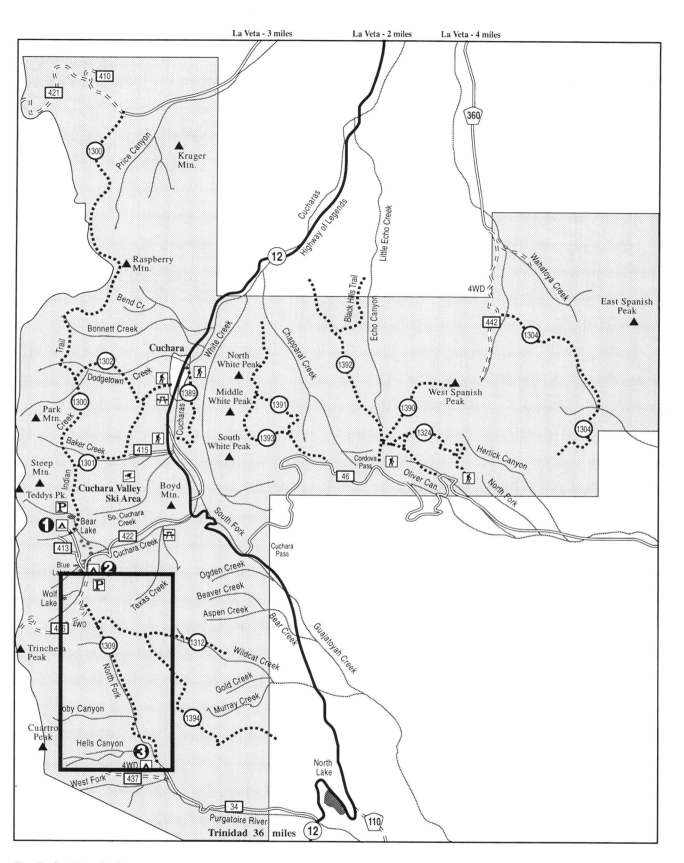

MAP 22
TRAIL #1309

© 1995 — *Outdoor Books & Maps, Inc.* • Denver, Colorado • (303) 629-6111

ROCKY MOUNTAIN REGION

MAP NUMBER(S): 20

NATIONAL FOREST: SAN ISABEL
RANGER DISTRICT: SAN CARLOS

USGS: RYE, HAYDEN BUTTE, BADITO CONE SAN ISABEL QUADS

DIFFICULTY: DIFFICULT

DISTANCE: 8.5 MILES

TRAIL 1310 BARTLETT

UPDATED:

TRAIL BEGINNING ELEVATION:
7,720 ft. FS Rd. #427

TRAIL ENDING ELEVATION:
11,600 ft. FS Rd. #369

USE: Moderate

RECOMMENDED SEASON:

| SPRING | SUMMER | FALL | WINTER |

ACCESS:
1.5 miles south of Rye to FSR #427

ATTRACTIONS:
Greenhorn Peak, scenery. Trail can be hot & dry, be sure to carry drinking water. Trail is in Greenhorn Mountain Wilderness.

NARRATIVE:

TRAIL PROFILE (ONE-WAY):

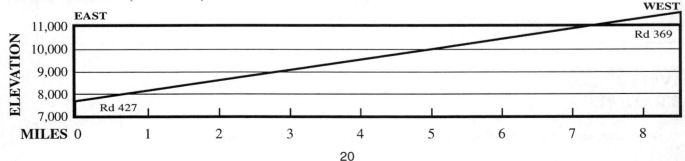

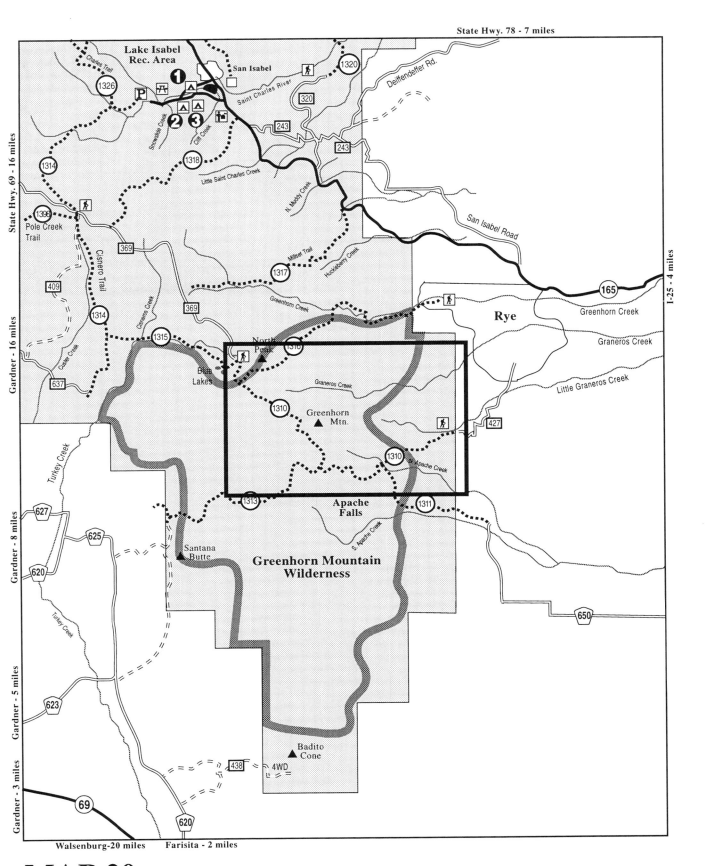

MAP 20
TRAIL #1310

© 1995 — Outdoor Books & Maps, Inc. • Denver, Colorado • (303) 629-6111

ROCKY MOUNTAIN REGION

MAP NUMBER(S): 22

NATIONAL FOREST: SAN ISABEL
RANGER DISTRICT: SAN CARLOS

USGS: TRINCHERA PEAK, CUCHARAS PASS QUADS

DIFFICULTY: MODERATE

DISTANCE: 3.0 MILES

TRAIL
1312
BAKER

UPDATED:

TRAIL BEGINNING ELEVATION:
1,400 ft. Intersection Trail #1309

TRAIL ENDING ELEVATION:
9,400 ft.

ACCESS:
18 miles from La Veta via Hwy 12, FS Rd. #422. Continue to intersection of Blue Lake Road #413 and Trinchera Peak Road #436; south on Trail #1309 to trailhead.

ATTRACTIONS:
Scenery

USE: Heavy

RECOMMENDED SEASON:

| SPRING | SUMMER | FALL | WINTER |

NARRATIVE:

TRAIL PROFILE (ONE-WAY):

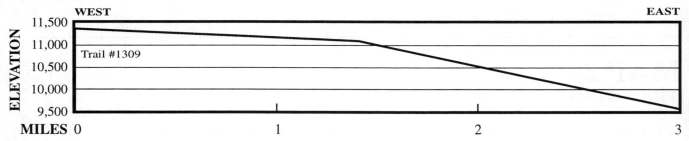

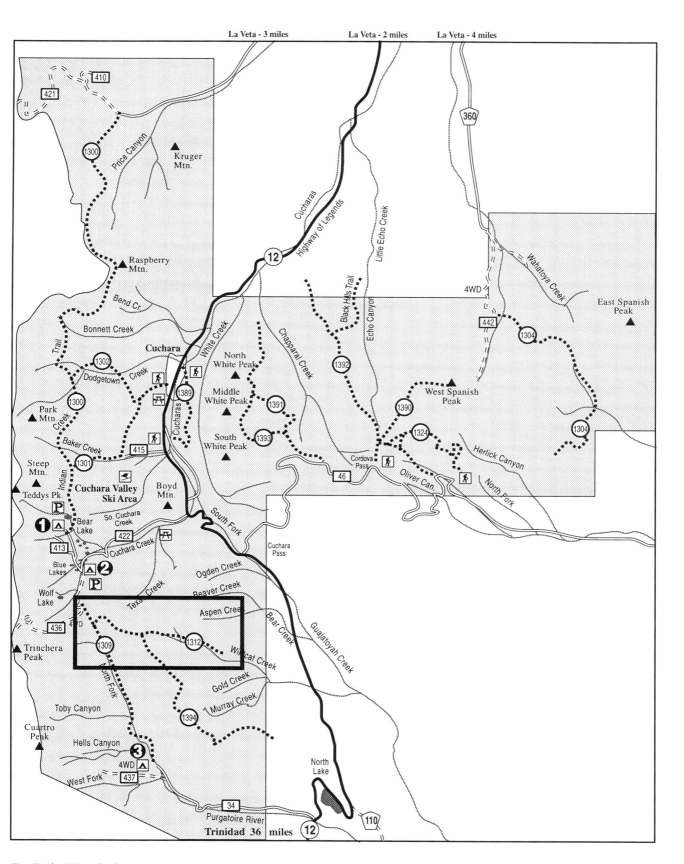

MAP 22
TRAIL #1312

© 1995 — *Outdoor Books & Maps, Inc.* • *Denver, Colorado* • **(303) 629-6111**

ROCKY MOUNTAIN REGION

MAP NUMBER(S): 20

NATIONAL FOREST: SAN ISABEL
RANGER DISTRICT: SAN CARLOS

USGS: SAN ISABEL QUAD

DIFFICULTY: MORE DIFFICULT, VERY STEEP
DISTANCE: 10.0 MILES

TRAIL 1314
CISNERO

UPDATED:

TRAIL BEGINNING ELEVATION:
9,000 ft. Parking lot of Lake Isabel Recreation Area

TRAIL ENDING ELEVATION:
9,180 ft. FS Rd. #637

ACCESS:
At Lake Isabel or off intersection of Forest Service Road #369 and Forest Service Road #409; or off FS Rd. #637.

ATTRACTIONS:
Scenery. Facilities at Lake Isabel Campground.

USE:

RECOMMENDED SEASON: SPRING | SUMMER | FALL | WINTER

TRAIL PROFILE (ONE-WAY):

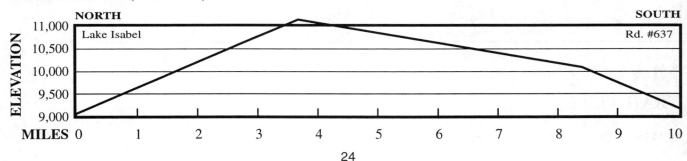

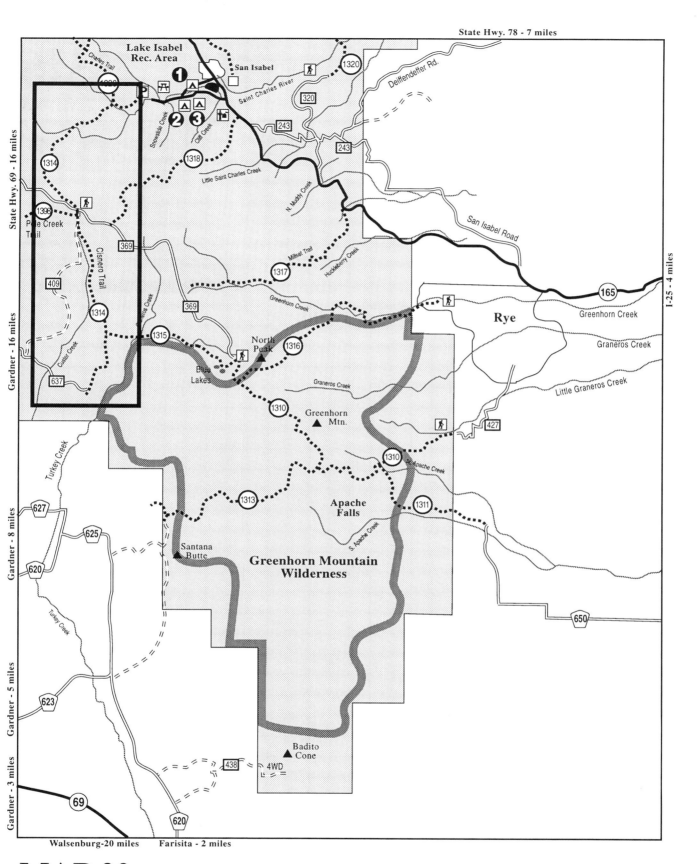

MAP 20
TRAIL #1314

© 1995 — *Outdoor Books & Maps, Inc.* • *Denver, Colorado* • **(303) 629-6111**

ROCKY MOUNTAIN REGION

MAP NUMBER(S): 20

NATIONAL FOREST: SAN ISABEL
RANGER DISTRICT: SAN CARLOS

USGS: RYE, SAN ISABEL QUADS

DIFFICULTY: DIFFICULT

DISTANCE: 9.5 MILES

TRAIL 1316 GREENHORN

UPDATED:

TRAIL BEGINNING ELEVATION:
7,450 ft. Cuerna Verde Park

TRAIL ENDING ELEVATION:
11,480 ft. Greenhorn Road

ACCESS:
From Rye City or FSR #369.

ATTRACTIONS:
North Peak, scenery.

USE: Moderate

RECOMMENDED SEASON:

| SPRING | SUMMER | FALL | WINTER |

NARRATIVE:

TRAIL PROFILE (ONE-WAY):

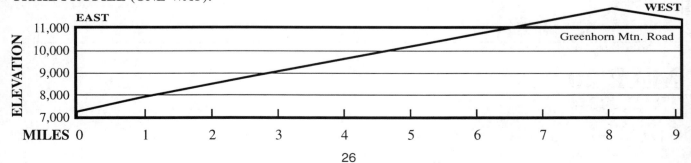

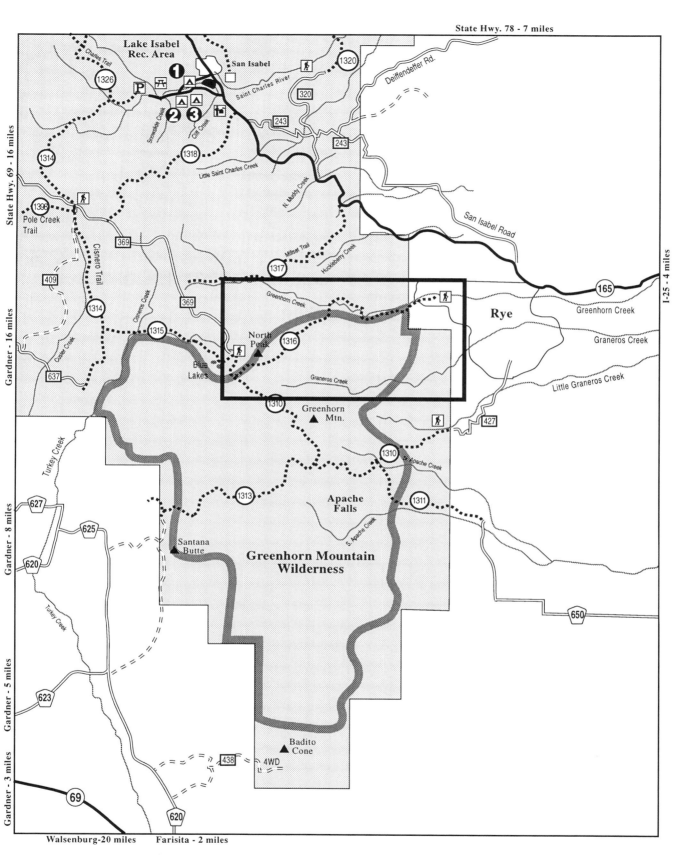

MAP 20
TRAIL #1316

© 1995 — *Outdoor Books & Maps, Inc.* • *Denver, Colorado* • **(303) 629-6111**

ROCKY MOUNTAIN REGION
MAP NUMBER(S): 20

NATIONAL FOREST: SAN ISABEL
RANGER DISTRICT: SAN CARLOS

USGS: SAN ISABEL QUAD

DIFFICULTY: MODERATE

DISTANCE: 6.0 MILES

TRAIL 1318 SNOWSLIDE

UPDATED:

TRAIL BEGINNING ELEVATION:
8,720 ft. Hwy #165

TRAIL ENDING ELEVATION:
11,480 ft. FS Rd. #369

ACCESS:
1 mile south of Lake Isabel via Hwy 165.

ATTRACTIONS:
Mountain scenery.

USE: Moderate

RECOMMENDED SEASON:
SPRING | SUMMER | FALL | WINTER

NARRATIVE:

TRAIL PROFILE (ONE-WAY):

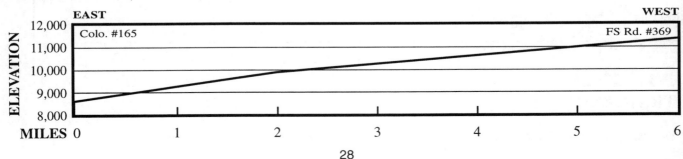

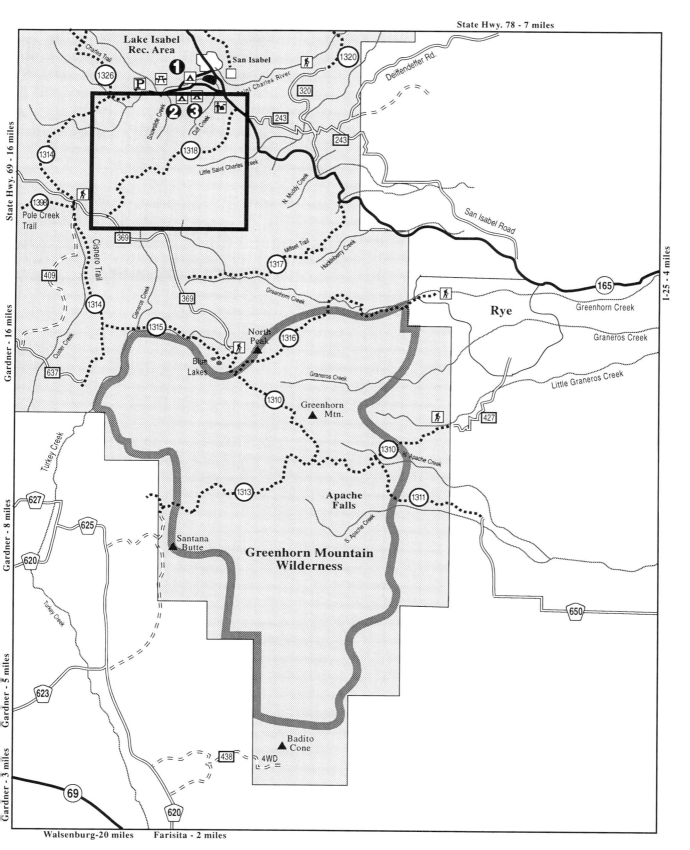

MAP 20
TRAIL #1318

© 1995 — *Outdoor Books & Maps, Inc.* • *Denver, Colorado* • (303) 629-6111

ROCKY MOUNTAIN REGION

MAP NUMBER(S): 17

NATIONAL FOREST: SAN ISABEL
RANGER DISTRICT: SAN CARLOS

USGS: ST. CHARLES PEAK QUAD

DIFFICULTY: MODERATE

DISTANCE: 5.5 MILES

TRAIL 1321
SOUTH CREEK

UPDATED: DECEMBER 1987

TRAIL BEGINNING ELEVATION:
9,120 ft. Hwy #165

TRAIL ENDING ELEVATION:
6,720 ft. Pueblo Mountain Park

ACCESS:
5 miles north of San Isabel, Highway #165.

ATTRACTIONS:
Scenery.

USE: Moderate

RECOMMENDED SEASON: SPRING | **SUMMER** | **FALL** | WINTER

NARRATIVE:

TRAIL PROFILE (ONE-WAY):

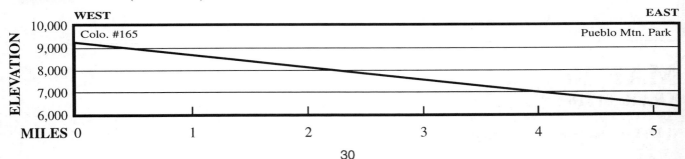

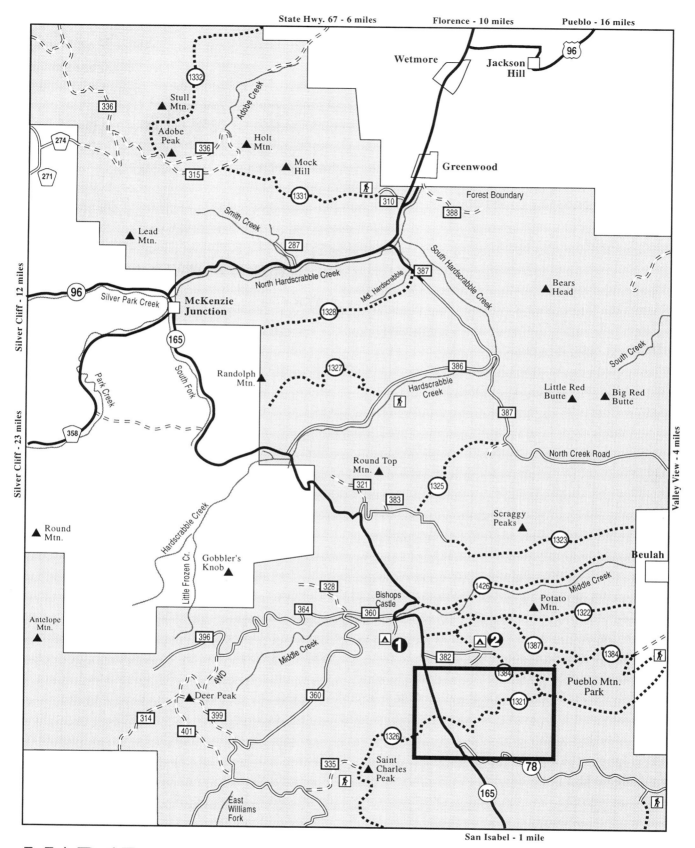

MAP 17
TRAIL #1321

© *1995 — Outdoor Books & Maps, Inc.* • *Denver, Colorado* • *(303) 629-6111*

ROCKY MOUNTAIN REGION
MAP NUMBER(S): 17

NATIONAL FOREST: SAN ISABEL
RANGER DISTRICT: SAN CARLOS

USGS: ST. CHARLES PEAK QUAD

DIFFICULTY: MODERATE

DISTANCE: 4.0 MILES

TRAIL 1322
SECOND MACE

UPDATED:

TRAIL BEGINNING ELEVATION:
9,000 ft. Hwy #165

TRAIL ENDING ELEVATION:
6,600 ft. N.F. Boundary west of Beulah

ACCESS:
6 miles north of Lake Isabel, Hwy #165

ATTRACTIONS:
Scenery

USE: Moderate

RECOMMENDED SEASON:

| SPRING | SUMMER | FALL | WINTER |

NARRATIVE:

TRAIL PROFILE (ONE-WAY):

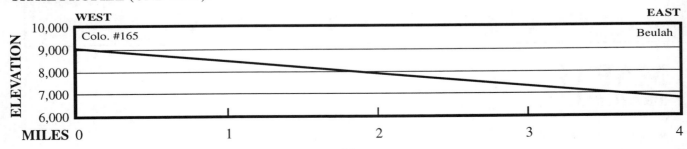

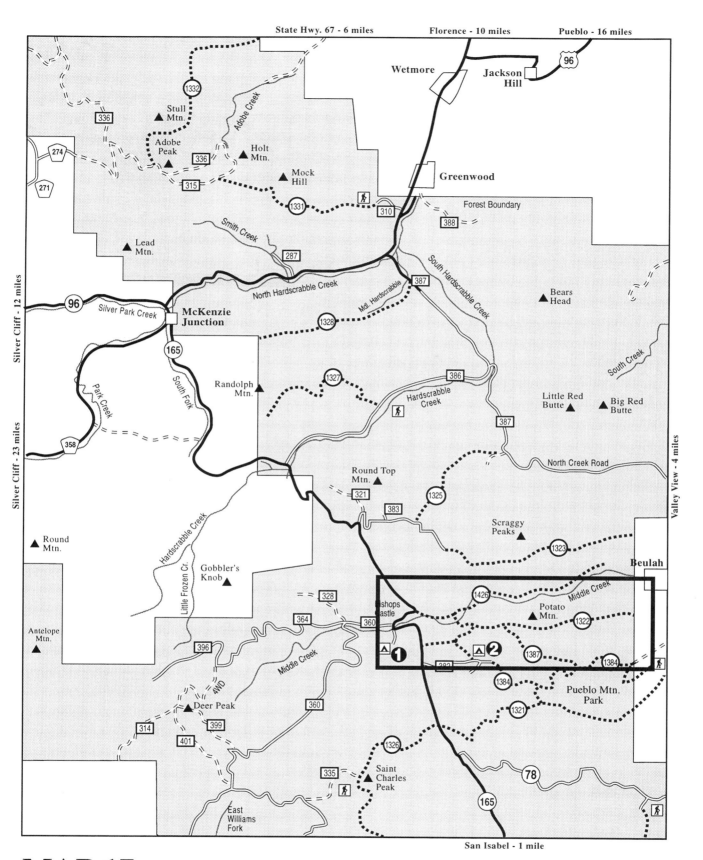

MAP 17
TRAIL #1322

© 1995 — *Outdoor Books & Maps, Inc.* • *Denver, Colorado* • (303) 629-6111

ROCKY MOUNTAIN REGION
MAP NUMBER(S): 17

NATIONAL FOREST: SAN ISABEL
RANGER DISTRICT: SAN CARLOS

USGS: ST. CHARLES PEAK QUAD

DIFFICULTY: MODERATE

DISTANCE: 8.0 MILES

TRAIL
1323
SILVER CIRCLE

UPDATED:

TRAIL BEGINNING ELEVATION:
9,400 ft. Silver Circle Trailhead on FDR #383

TRAIL ENDING ELEVATION:
6,800 ft. N.F. Boundary Panther Creek

ACCESS:
10 miles north of San Isabel, Hwy #165 to FSR #321

ATTRACTIONS:
Scenery, carry water, facilities at Bigelow Trailhead

USE: Moderate

RECOMMENDED SEASON:

| SPRING | SUMMER | FALL | WINTER |

NARRATIVE:

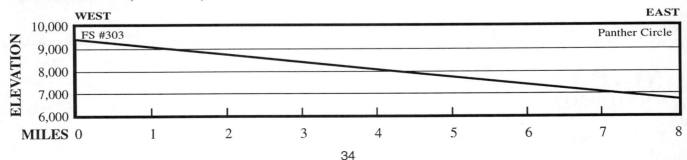

TRAIL PROFILE (ONE-WAY):

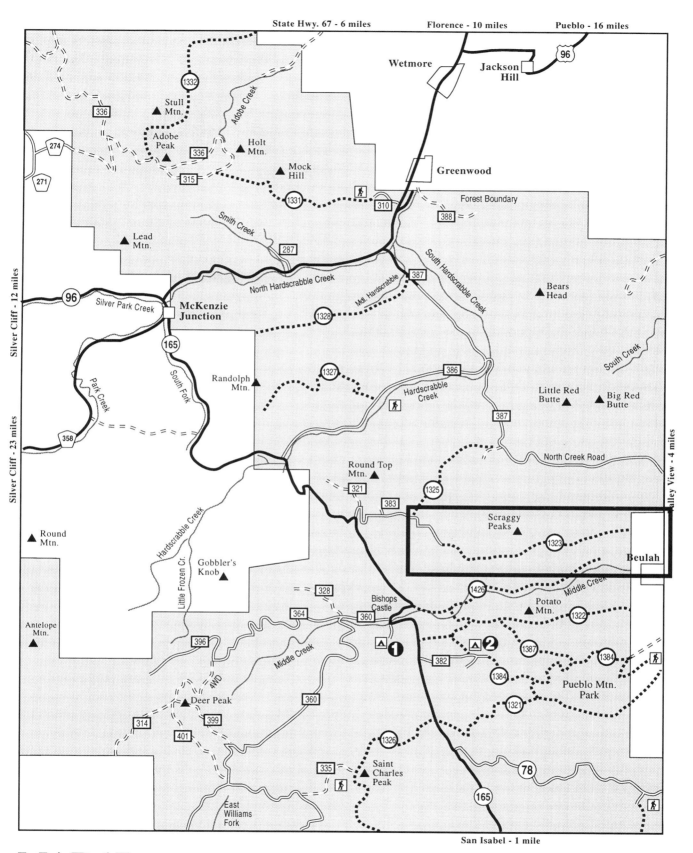

MAP 17
TRAIL #1323

© 1995 — *Outdoor Books & Maps, Inc.* • *Denver, Colorado* • (303) 629-6111

ROCKY MOUNTAIN REGION

MAP NUMBER(S): 17

NATIONAL FOREST: SAN ISABEL
RANGER DISTRICT: SAN CARLOS

USGS: ST. CHARLES PEAK QUAD

DIFFICULTY: MODERATE

DISTANCE: 3.5 MILES

TRAIL
1325
NORTH CREEK

UPDATED:

TRAIL BEGINNING ELEVATION:
7,600 ft. FS Rd. #387

TRAIL ENDING ELEVATION:
9,200 ft. FS Rd. #383

ACCESS:
10 miles North of Lake Isabel via Hwy 154 to Bigelow Divide Road #321 and Silver Circle Trail (No public access thru private land located next to North Creek Road #307).

ATTRACTIONS:
Mountain scenery.

USE: Moderate

RECOMMENDED SEASON:

| SPRING | SUMMER | FALL | WINTER |

NARRATIVE:

TRAIL PROFILE (ONE-WAY):

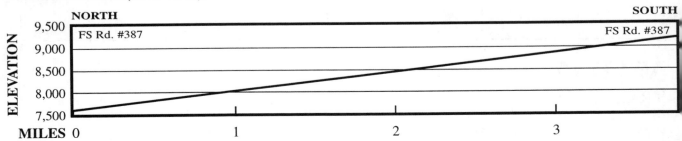

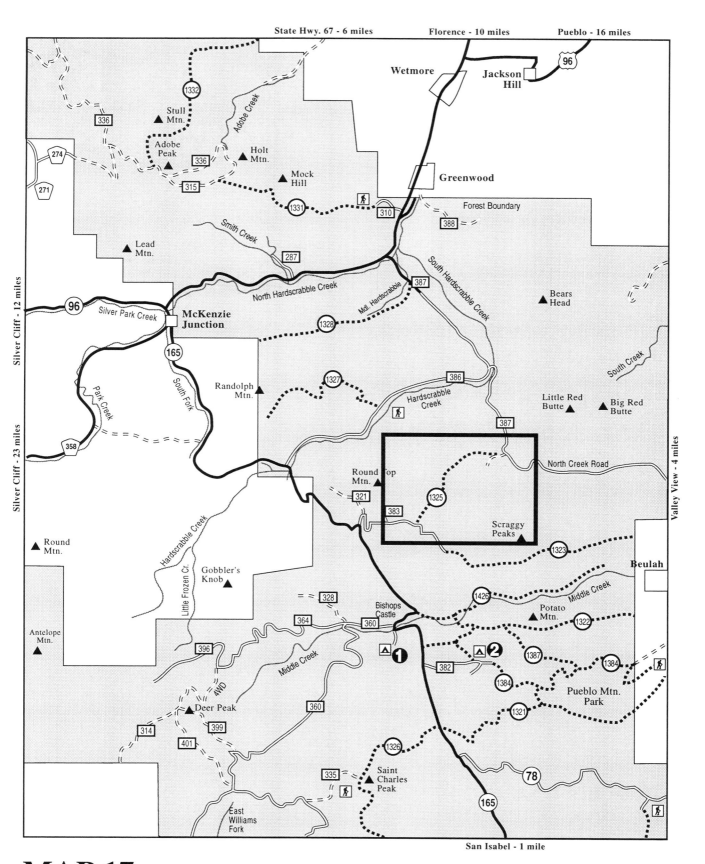

MAP 17
TRAIL #1325

© 1995 — Outdoor Books & Maps, Inc. • Denver, Colorado • (303) 629-6111

ROCKY MOUNTAIN REGION

MAP NUMBER(S): 17 and 20

NATIONAL FOREST: SAN ISABEL
RANGER DISTRICT: SAN CARLOS

USGS: ST. CHARLES PEAK, SAN ISABEL QUADS

DIFFICULTY: DIFFICULT

DISTANCE: 7.5 MILES

TRAIL 1326
SAINT CHARLES

UPDATED:

TRAIL BEGINNING ELEVATION:
8,920 ft. Lake Isabel

TRAIL ENDING ELEVATION:
9,080 ft. Hwy #165 - Greenhill Divide

ACCESS:
At Lake Isabel Rest Area Hwy #165 or Hwy #165 at Greenhill Divide.

ATTRACTIONS:
St Charles Peak, scenery, facilities at Lake Isabel Campground.
No horse or motorbike access from Lake Isabel Rest Area.

USE: Moderate

RECOMMENDED SEASON:

| SPRING | SUMMER | FALL | WINTER |

NARRATIVE:

TRAIL PROFILE (ONE-WAY):

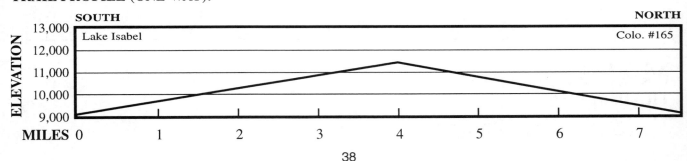

MAP 17 - South East Half

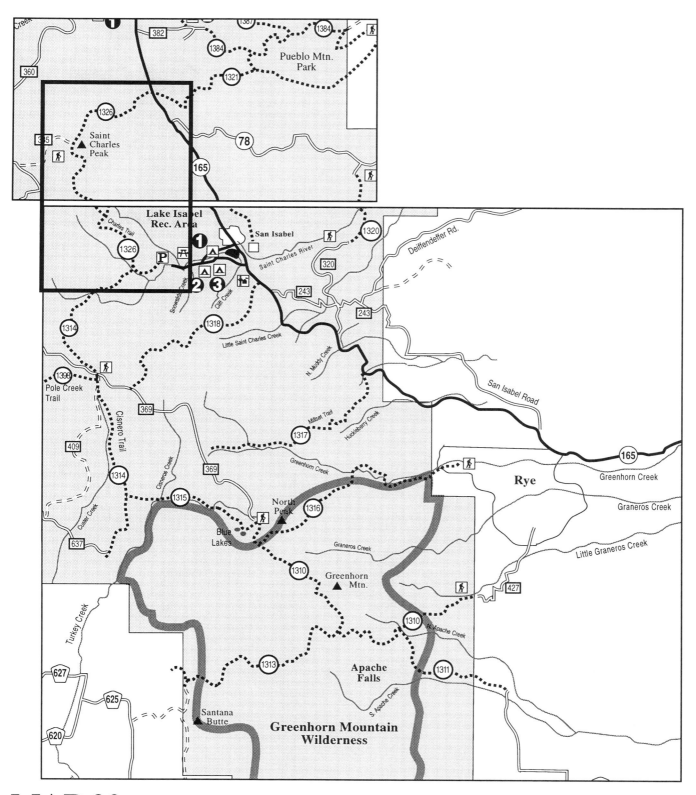

MAP 20 - North West Half
TRAIL #1326

© *1995 — Outdoor Books & Maps, Inc.* • Denver, Colorado • **(303) 629-6111**

ROCKY MOUNTAIN REGION

MAP NUMBER(S): 17

NATIONAL FOREST: SAN ISABEL
RANGER DISTRICT: SAN CARLOS

USGS: HARDSCRABBLE MOUNTAIN, WETMORE QUADS

DIFFICULTY: MODERATE

DISTANCE: 4.5 MILES

TRAIL 1327
RUDOLPH MOUNTAIN

UPDATED:

TRAIL BEGINNING ELEVATION:
8,000 ft. S. Hardscrabble Road

TRAIL ENDING ELEVATION:
9,280 ft. Private Land near Hwy #165

ACCESS:
5 miles south of Wetmore on Hwy 165 to FS Rd #387 south to FS Rd #386, east to trailhead.

ATTRACTIONS:
Scenery, carry water, facilities at Florence Picnic Grounds
Trail is difficult for trailbikes.

USE: Moderate

RECOMMENDED SEASON:

| SPRING | SUMMER | FALL | WINTER |

NARRATIVE:

TRAIL PROFILE (ONE-WAY):

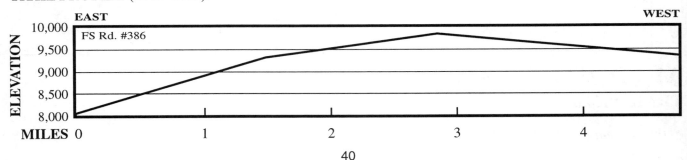

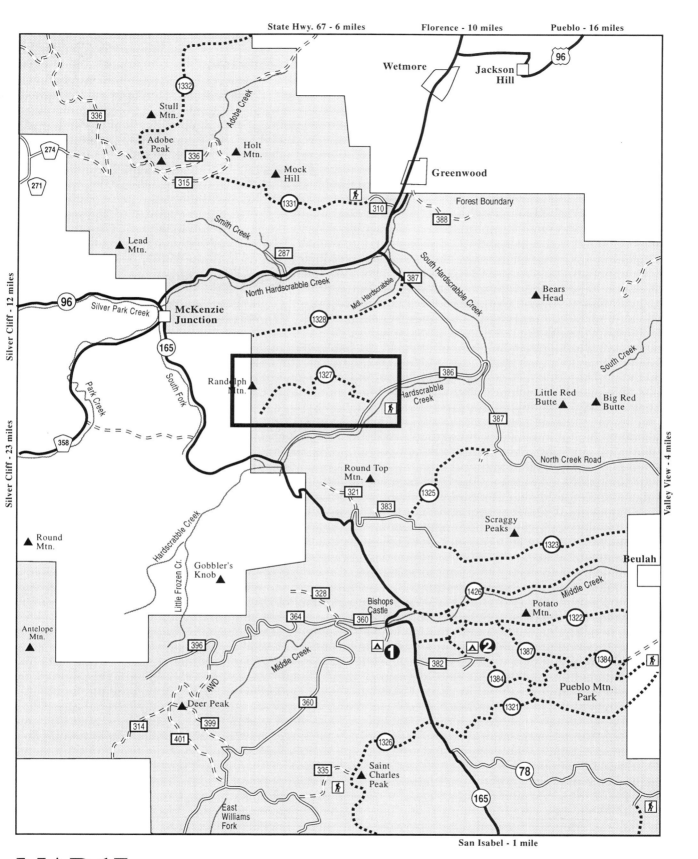

MAP 17
TRAIL #1327

© 1995 — *Outdoor Books & Maps, Inc.* • *Denver, Colorado* • (303) 629-6111

ROCKY MOUNTAIN REGION
MAP NUMBER(S): 17

NATIONAL FOREST: SAN ISABEL
RANGER DISTRICT: SAN CARLOS

USGS: HARDSCRABBLE MOUNTAIN, WETMORE QUADS

UPDATED:

TRAIL BEGINNING ELEVATION:
6,690 ft. Hwy #96

TRAIL ENDING ELEVATION:
9,840 ft. FSR #315

ACCESS:
3 miles south of Wetmore, Hwy #96.

ATTRACTIONS:
Scenery, carry water, no facilities, First 1/2 mile of trail crosses private land. Stay on public right-of-way across this area. Trail is difficult for trailbikes.

DIFFICULTY: MODERATE

DISTANCE: 5.0 MILES

TRAIL 1331
LEWIS CREEK

USE: Moderate

RECOMMENDED SEASON:

| SPRING | SUMMER | FALL | WINTER |

NARRATIVE:

TRAIL PROFILE (ONE-WAY):

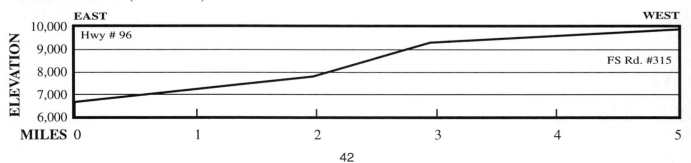

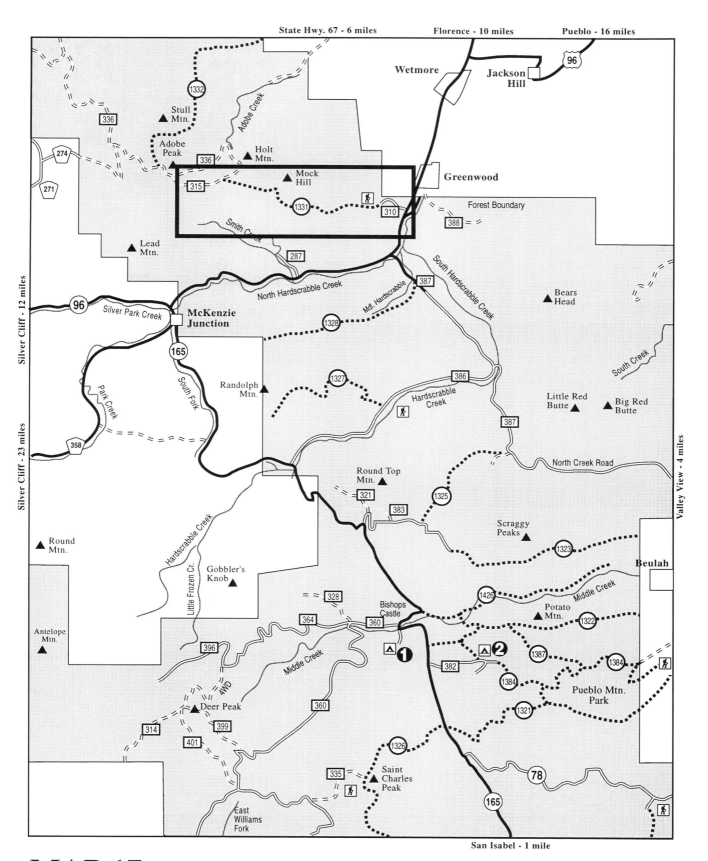

MAP 17
TRAIL #1331

© 1995 — *Outdoor Books & Maps, Inc.* • *Denver, Colorado* • (303) 629-6111

ROCKY MOUNTAIN REGION

MAP NUMBER(S): 15 and 17

NATIONAL FOREST: SAN ISABEL
RANGER DISTRICT: SAN CARLOS

USGS: ROCKVALE, HARDSCRABBLE MOUNTAIN QUADS

UPDATED:

TRAIL BEGINNING ELEVATION:
7,000 ft. Mineral Creek

TRAIL ENDING ELEVATION:
10,188 ft. Adobe Peak

ACCESS:
20 miles south of Canon City, County Road #143 to #274; to FDR #336.

ATTRACTIONS:
Scenery, carry water, trail difficult for trailbikes

DIFFICULTY: MODERATE

DISTANCE: 4.5 MILES

TRAIL 1332
MINERAL-STEVENS

USE: Moderate

RECOMMENDED SEASON:

| SPRING | SUMMER | FALL | WINTER |

NARRATIVE:

TRAIL PROFILE (ONE-WAY):

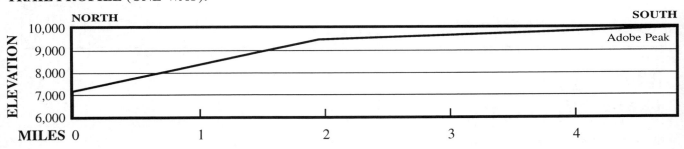

MAP 15 - South Half

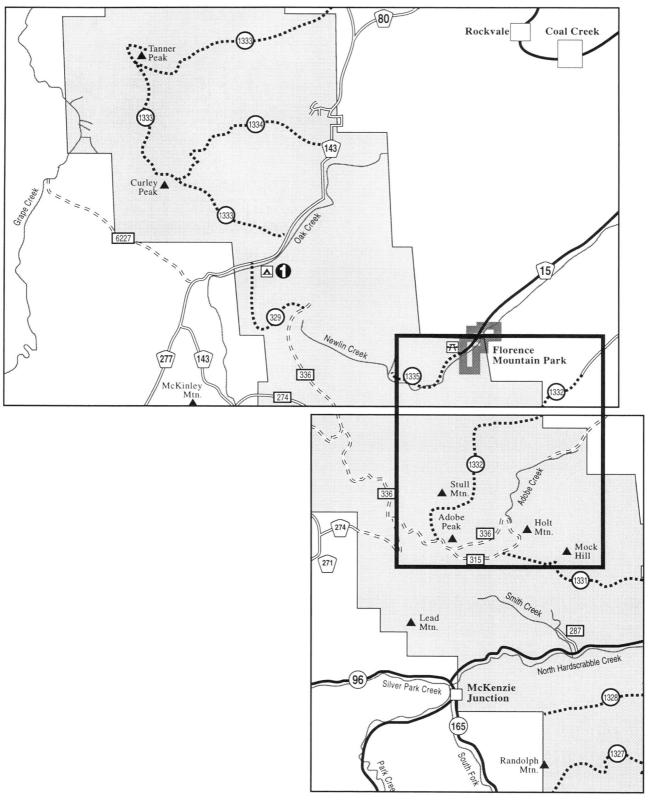

MAP 17 - North West Qtr.

TRAIL #1332

© *1995 — Outdoor Books & Maps, Inc.* • *Denver, Colorado* • **(303) 629-6111**

ROCKY MOUNTAIN REGION

MAP NUMBER(S): 15

NATIONAL FOREST: SAN ISABEL
RANGER DISTRICT: SAN CARLOS

USGS: CURLEY PEAK QUAD

DIFFICULTY: MODERATE

DISTANCE: 15.5 MILES

TRAIL 1333
TANNER

UPDATED:

TRAIL BEGINNING ELEVATION:
7,360 ft. E. Bear Gulch on County Rd #143

TRAIL ENDING ELEVATION:
6,000 ft. Cow Circle on County Rd #143

ACCESS:
4 miles south of Canon City, County Road #143, and 10 miles south of Canon City, County #143.

ATTRACTIONS:
Scenery, no live streams, must carry water, trail if difficult for trail bikes.

USE: Moderate

RECOMMENDED SEASON:

| SPRING | SUMMER | FALL | WINTER |

NARRATIVE:

TRAIL PROFILE (ONE-WAY):

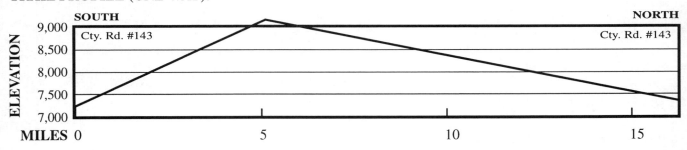

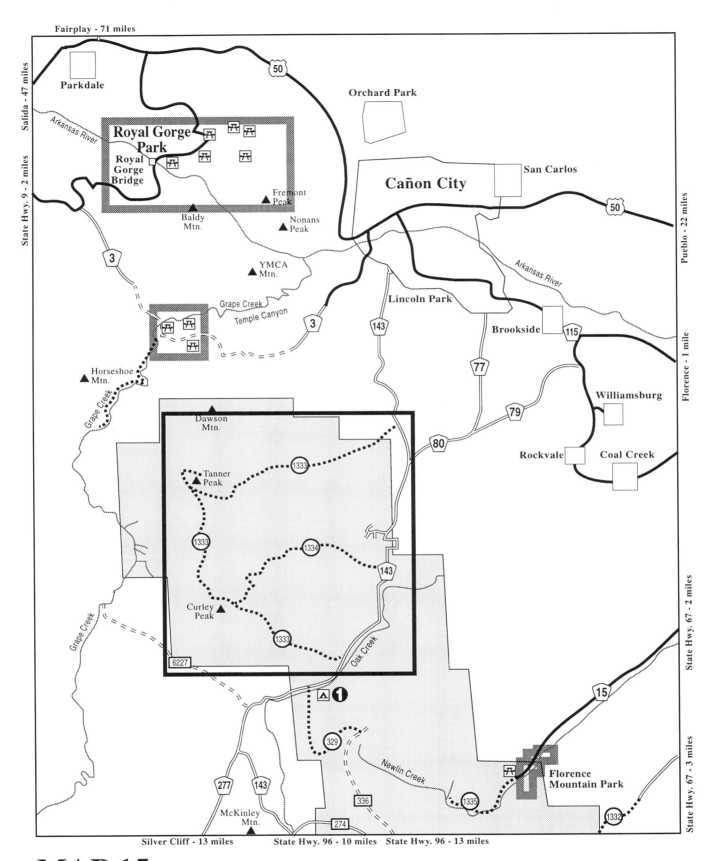

MAP 15
TRAIL #1333

© *1995 — Outdoor Books & Maps, Inc.* • *Denver, Colorado* • **(303) 629-6111**

ROCKY MOUNTAIN REGION

MAP NUMBER(S): 18 and 16

NATIONAL FOREST: SAN ISABEL
RANGER DISTRICT: SAN CARLOS

USGS: HORN PEAK, CRESTONE PEAK, BECK MOUNTAIN QUADS

UPDATED: NOVEMBER 1987

TRAIL BEGINNING ELEVATION:
9,260 ft. Grape Creek at Music Pass Road

TRAIL ENDING ELEVATION:
9,430 ft. Horn Creek

TRAIL 1336A-B
GRAPE CREEK-HORN CREEK

DIFFICULTY: EASY TO MODERATE
DISTANCE: 14.5 MILES

USE: Moderate

RECOMMENDED SEASON: SPRING | **SUMMER** | **FALL** | WINTER

ACCESS:
Horn Creek: From Westcliffe, take the Hermit Lake Road one mile west to Macey Lane. There head south five miles to Horn Road. Then three miles west and south to Rainbow Trail. Stay on road through private land at Horn Creek Ranch.

Music Pass Road: From Westcliffe go south 4 miles on highway 69 to Colfax Lane. Then turn right and follow this road south 5.5 miles to the crossroad. Turn left 1/4 mile and south 5 miles to National Forest land and trailhead at Grape Creek.

South Colony Creek: From Westcliffe, go south as described above to the crossroad. There turn right instead. At about 2 miles the road becomes a four-wheel drive road climbing steeply another mile to the trail crossing and beyond up South Colony Creek.

ATTRACTIONS:
Hikers and trail bikers use this section for access to the trails to the many lakes and outstanding scenery of the high country of the Sangre de Cristo Range. Fishing is popular in the streams as well as in the high mountain lakes such as Hermit Lake, Venable Lakes or Comanche Lake to name a few. State fishing laws apply. Licenses required. Side trails are plentiful to the above lakes and other destinations. Some offer access to the west side of the Sangre de Cristo Range.

NARRATIVE:
Trails OPEN to motorized vehicles: #1336 Rainbow, #329 Music Pass Road and #313 South Colony Road.

Trails CLOSED to motorized vehicles: #1337 Music Pass, #1338 Hudson Creek, #1339 South Colony Lakes, #1340 North Colony, #1341 Macey Creek, #1342 Horn Creek and on the Rio Grande National Forest #743 Sand Creek. Visitors are urged to respect the use of others for mutual safety and enjoyment

TRAIL PROFILE (ONE-WAY):

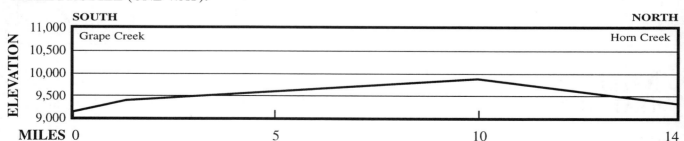

MAP 16 - South East Quarter

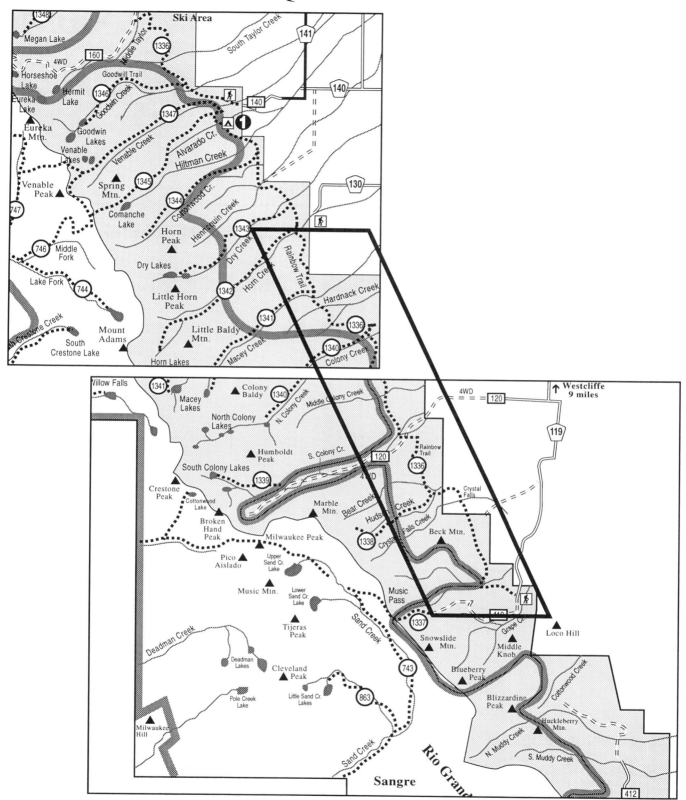

MAP 18 - North Half
TRAIL #1336A-B

© 1995 — *Outdoor Books & Maps, Inc.* • *Denver, Colorado* • (303) 629-6111

ROCKY MOUNTAIN REGION
MAP NUMBER(S): 16

NATIONAL FOREST: SAN ISABEL
RANGER DISTRICT: SAN CARLOS

USGS: HORN PEAK QUAD

DIFFICULTY: EASY TO MODERATE
DISTANCE: 7.5 MILES

TRAIL 1336B-C
HORN CREEK-MIDDLE TAYLOR CREEK

UPDATED: NOVEMBER 1987

TRAIL BEGINNING ELEVATION:
9,430 ft. Horn Creek

TRAIL ENDING ELEVATION:
9,960 ft. Middle Taylor Creek

USE: Heavy

RECOMMENDED SEASON:

| SPRING | SUMMER | FALL | WINTER |

ACCESS:
Horn Creek: From Westcliffe, take the Hermit Lake Road one mile west to Macey Lane. There head south five miles to Horn Road. Then three miles west and south to Rainbow Trail. Stay on road through private land at Horn Creek Ranch.
Middle Taylor Creek: Take Hermit Lake road #301 about 8 miles west of Westcliffe. From the primitive campsite the trail follows the road to the north about 1 miles downstream to where it continues northward. There are no facilities at this site.
Alvarado Campground: From Westcliffe, take the Hermit Lake Road 1 mile west to Macey Lane, then 3 miles south to Schoolfield Road, then about 4 miles west and south to Alvarado Campground.

ATTRACTIONS:
Sections of this segment of the Rainbow Trail are the heaviest used. Hikers and trail bikers use this section for access to the trails to the many lakes and outstanding scenery of the high country of the Sangre de Cristo Range. Fishing is popular in the streams as well as in the high mountain lakes such as Hermit Lake, Venable Lakes or Comanche Lake to name a few. State fishing laws apply. Licenses required. Camping is allowed in undeveloped sites along the trail. Please leave a clean camp and dead fire. Alvarado Campground is an excellent developed campground. A user fee is charged at this site.

NARRATIVE:
Trails OPEN to motorized vehicles: #1336 Rainbow, #1345 Comanche, Venable except Phantom Terrace), #301 Hermit Pass Road.
Trails CLOSED to motor vehicles: #1342 Horn Creek, #1343 Dry Creek, #1`344 Cottonwood, #1345 Phantom Terrace, #1346 Goodwin. On the Rio Grande National Forest trails No's: #744, 745,746,747,859 and 1345.

TRAIL PROFILE (ONE-WAY):

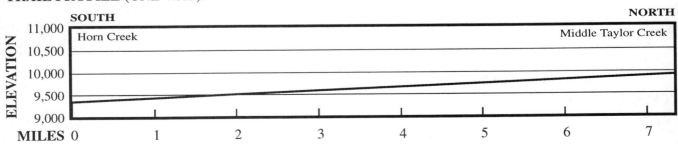

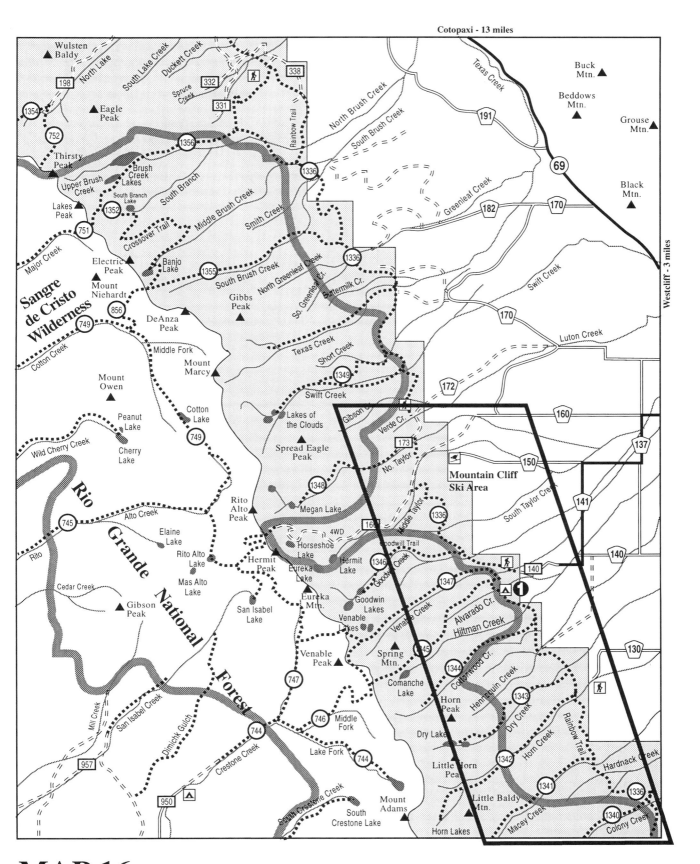

MAP 16
TRAIL #1336B-C

ROCKY MOUNTAIN REGION

MAP NUMBER(S): 16

NATIONAL FOREST: SAN ISABEL
RANGER DISTRICT: SAN CARLOS

USGS: BECKWITH MOUNTAIN, ELECTRIC PEAK, HORN PEAK QUADS

DIFFICULTY: MODERATE

DISTANCE: 11.0 MILES

TRAIL 1336C-D

MIDDLE TAYLOR CREEK-N. BRUSH CREEK

UPDATED: NOVEMBER 1987

TRAIL BEGINNING ELEVATION:
9,960 ft. Middle Taylor Creek

TRAIL ENDING ELEVATION:
8,820 ft. North Brush Creek

USE: Moderate

RECOMMENDED SEASON: SPRING, **SUMMER**, **FALL**, WINTER

ACCESS:
North Brush Creek: The trailhead provides access from road #337 located about 4 miles southwest of Hillside by way of forest roads #300 and 332. No water or facilities are provided.
Middle Taylor Creek: Take Hermit Lake road #301 about 8 miles west of Westcliffe. From the primitive campsite the trail follows the road to the north about one mile downstream to where it continues northward. There are no facilities at this site. Other roads such as the Pines or Greenleaf Creek pass through private lands not open to the public.

ATTRACTIONS:
Undeveloped camping sites along the trail and at the trailheads are available. No facilities or drinking water are provided. Campfire permits are not required but please make sure your fire is dead out before you leave. No trash pickup is provided. Fishing is popular in the streams as well as in the nearby Eureka Lake, Hermit Lake, Lakes of the Clouds, and Brush Creek Lakes. State fishing laws apply. Licenses are required. Side trails are plentiful to the above lakes and other destinations. Some offer access to the west side of the Sangre de Cristo Range.

NARRATIVE:
Trails OPEN to motorized vehicles: #1336 Rainbow, #301 4-Wheel Drive road to Hermit Pass, #1349 Lakes of the Clouds.
Trails CLOSED to motorized vehicles: #751 North Brush Creek, #856 South Brush Creek, #1348 North Taylor, #1350 Texas Creek, #1352 Crossover, #1353 North Brush, and on the Rio Grande National Forest trails No's. #745, #749 and #751.

TRAIL PROFILE (ONE-WAY):

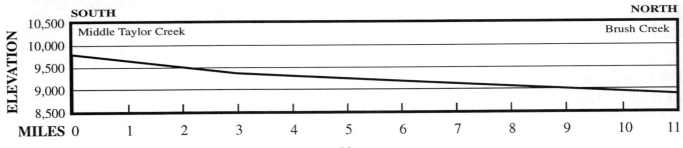

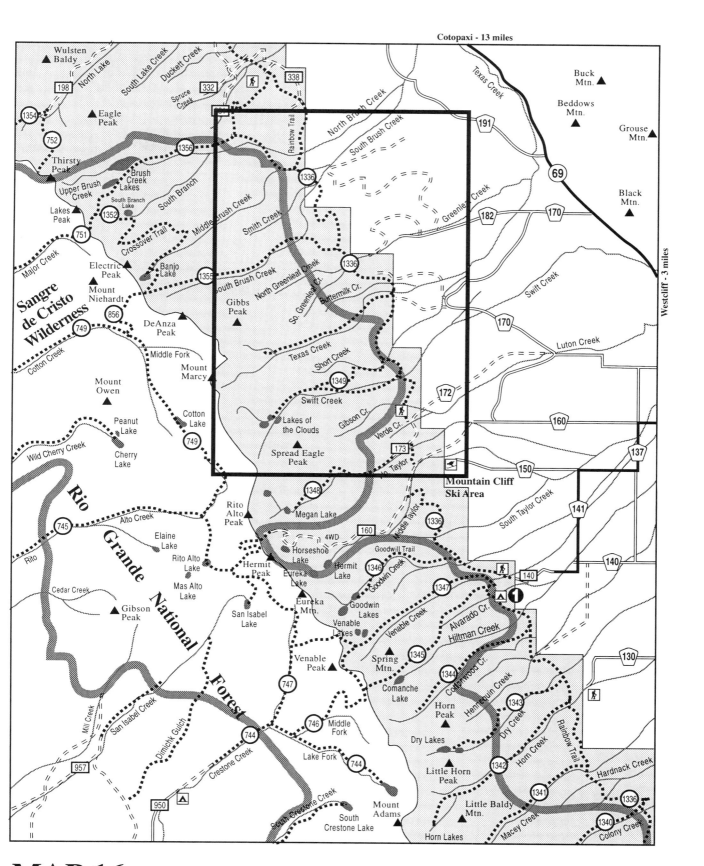

MAP 16
TRAIL #1336C-D

© 1995 — *Outdoor Books & Maps, Inc.* • *Denver, Colorado* • (303) 629-6111

ROCKY MOUNTAIN REGION

MAP NUMBER(S): 16 and 14

NATIONAL FOREST: SAN ISABEL
RANGER DISTRICT: SALIDA

USGS: ELECTRIC PEAK, COTOPAXI QUADS

DIFFICULTY: EASY

DISTANCE: 12.0 MILES

TRAIL
1336D-E

N. BRUSH CREEK-BIG COTTONWOOD CREEK

UPDATED: NOVEMBER 1987

TRAIL BEGINNING ELEVATION:
8,820 ft. North Brush Creek

TRAIL ENDING ELEVATION:
7,320 ft. Big Cottonwood Creek

USE: Light

RECOMMENDED SEASON:

| SPRING | SUMMER | FALL | WINTER |

ACCESS:
North Brush Creek: The Brush Creek Trailhead provides access from Road No. 337 located about 4 miles southwest of Hillside by way of Forest Rds No. 300 and 332 as shown on the map. No water or facilities are provided.
Balman Reservoir Road: The Rainbow trail crosses and follows Forest Rd 300 for a short distance. No facilities or parking are provided at this site.
Big Cottonwood Creek: At 1 mile southwest of Coaldale, take County Rd No. 140 about 3-1/2 miles to end. No drinking water or facilities are provided.
The above sites are accessible by autos and small trucks with small trailers.

ATTRACTIONS:
Undeveloped sites along the trail and at the Trailheads are available for camping. No facilities or drinking water are provided. Campfire permits are not required but please drown your campfire and make sure it is dead out before you leave it. No trash pickup is provided at any of these locations so please pack it out with you. Do not try to bury your trash as the animals will dig it up later. Drinking water from streams along the way should be treated
before using. Fishing is popular from the side trails along the route, particularly at Brush Creek, Rainbow and Balman Reservoir Lakes. State Regulations apply.

NARRATIVE:
Trails OPEN to Motorized Trail Vehicle
1336 - Rainbow Trail
300 - 4 WD road to Rainbow Lake

Trails CLOSED to Motorized Trail Vehicle
751 - North Brush Creek Trail
752 - North Lake Creek Trail
1353 - North Brush Creek Access
1354 - Silver Lake Trail Please see map on back

TRAIL BEGINS: (This Segment) At North Brush Creek
Elev. 8,820 ft. (2690 m)

TRAIL ENDS: (This Segment)
At Big Cottonwood Creek Elev. 7,320 ft. (2236 m)

TRAIL PROFILE (ONE-WAY):

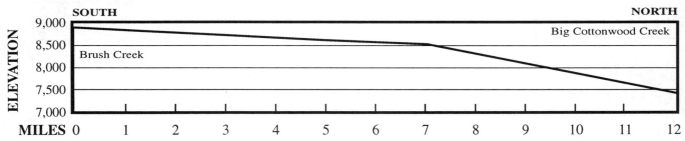

MAP 14 - South Half

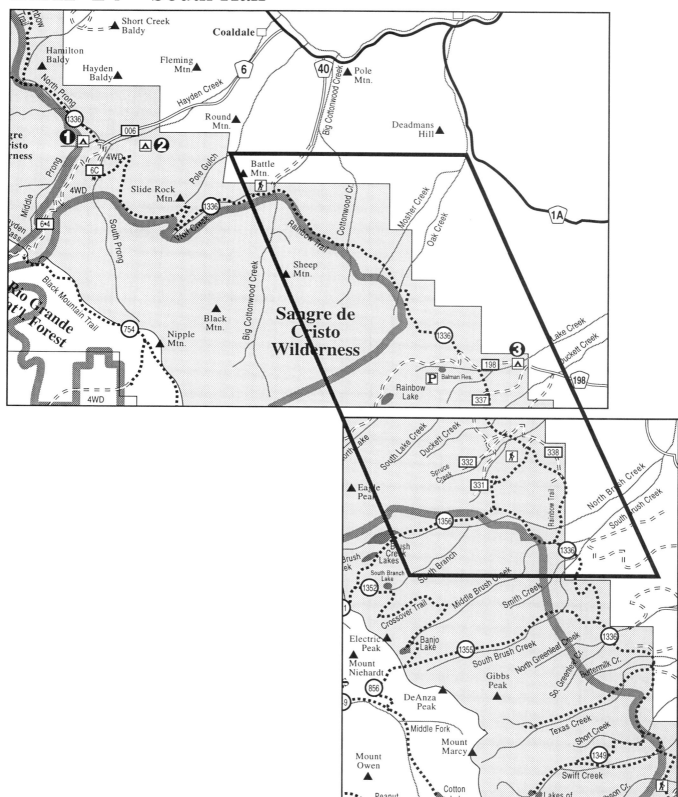

MAP 16 - North West Quarter
TRAIL #1336D-E

© *1995 — Outdoor Books & Maps, Inc.* • *Denver, Colorado* • (303) 629-6111

ROCKY MOUNTAIN REGION

MAP NUMBER(S): 14

NATIONAL FOREST: SAN ISABEL
RANGER DISTRICT: SALIDA

USGS: COALDALE, HOWARD QUADS

DIFFICULTY: MODERATE

DISTANCE: 13.5 MILES

TRAIL 1336E-F

BIG COTTONWOOD CREEK-STOUT CREEK

UPDATED: NOVEMBER 1987

TRAIL BEGINNING ELEVATION:
7,320 ft. Big Cottonwood Creek

TRAIL ENDING ELEVATION:
8,560 ft. Stout Creek

USE: Light

RECOMMENDED SEASON:

SPRING	SUMMER	FALL	WINTER

ACCESS:

Kerr Gulch: The Rainbow Trail at Stout Creek is reached from the end of the Kerr Gulch Rd. From US Hwy 50, 4 miles NW of Coaldale, take Kerr Gulch road BLM Rd. 6110 about 4 miles to BLM rd 6117. Then take Rd 6117 about 1 mile to end at the trailhead. This access is recommended for high clearance vehicles and dry weather only.

Hayden Creek: The trail passes through Hayden Creek Campground. From Coaldale take County Rd #6 about 5 miles to the campground.

Big Cottonwood: At 1 mile southwest of Coaldale, take County Rd #40 about 3.5 miles to end.

ATTRACTIONS:

Camping: Hayden Creek has 2 campgrounds, Coaldale Campground and Hayden Creek Campground both have toilets, fireplaces and tables. Only Hayden Creek campground has drinking eater.

Big Cottonwood Creek does not have a developed campground although there are good undeveloped sites there. Trash pickup service is not offered at these locations. Please pack your trash back home with you.

Hayden Pass was a short cut to the San Luis Valley during the 1800's. It is now a locally popular 4-Wheel drive road over the Sangre de Cristo Range. Although steep, narrow and rough, it offers excellent views of the area.

This section of the trail is open to motorized trail vehicles. Users are reminded to exercise caution and respect the rights and safety of other users. The side trails to Stout Creek Lakes, Bushnell Lakes, and up Big Cottonwood Creek are all closed to motor vehicles. Please be careful with fire. Drown your campfire to be sure it is dead out.

NARRATIVE:

TRAIL PROFILE (ONE-WAY):

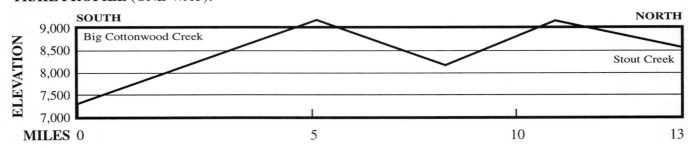

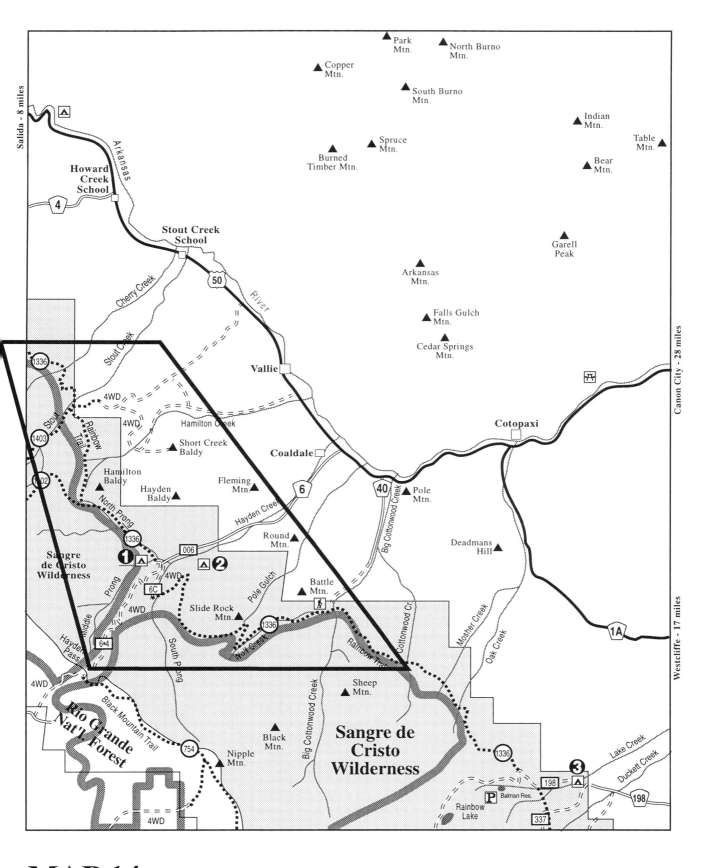

MAP 14
TRAIL #1336E-F

ROCKY MOUNTAIN REGION

MAP NUMBER(S): 13 AND 14

NATIONAL FOREST: SAN ISABEL
RANGER DISTRICT: SALIDA

USGS: HOWARD, WELLSVILLE QUADS

DIFFICULTY: MODERATE

DISTANCE: 12.2 MILES

TRAIL
1336F-G
STOUT CREEK-BEAR CREEK

UPDATED: OCTOBER 1987

TRAIL BEGINNING ELEVATION:
8,560 ft. Stout Creek

TRAIL ENDING ELEVATION:
8,880 ft. Bear Creek

USE: Light

RECOMMENDED SEASON:

SPRING	SUMMER	FALL	WINTER

ACCESS:

Kerr Gulch: The Rainbow Trail at Stout Creek is reached from the end of the Kerr Gulch Rd. From US Hwy 50, 4 mi NW of Coaldale, take Kerr Gl. Rd. (BLM Rd. No. 6110) About 4 miles to BLM Rd 6117. then take Rd 6117 about 1 mile to end at the trailhead. This access is recommended for high clearance vehicles and dry weather only.

Bear Creek: The trail is accessible from the end of Forest and County Road No.101, from US Hwy 50 about 2 miles southwest of Salida. The road is rough and recommended for high clearance vehicles only.

ATTRACTIONS:

Camping is permitted at undeveloped sites along the trail except at West Creek or Howard Creek where the Trail Right-of-Way passes through privately owned lands. The trailhead at the Bear Creek Road offers good undeveloped campsites. You are requested to pack out all of your trash however since no trash pickup service is offered at these isolated undeveloped locations. Other roads or trails such as Howard Creek, Porter Gulch, Spring Creek, West Creek or Cherry Creek which access the trail pass through private land which is not open to public use. Respect all posted lands. Do Not Trespass.

This section of the trail is open to motorized trail vehicles. Users are reminded to exercise caution and respect the rights and safety of other users. Although ATV's under 40 inches in width are allowed, the trail is not designed to accomodate most three or four wheel type vehicles. We recommend that you do not attempt to use them on the Rainbow trail.

NARRATIVE:

TRAIL PROFILE (ONE-WAY):

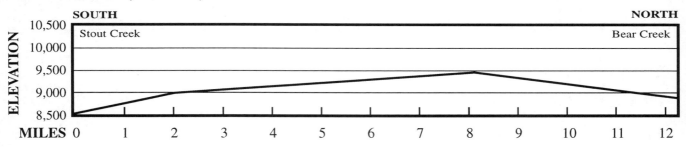

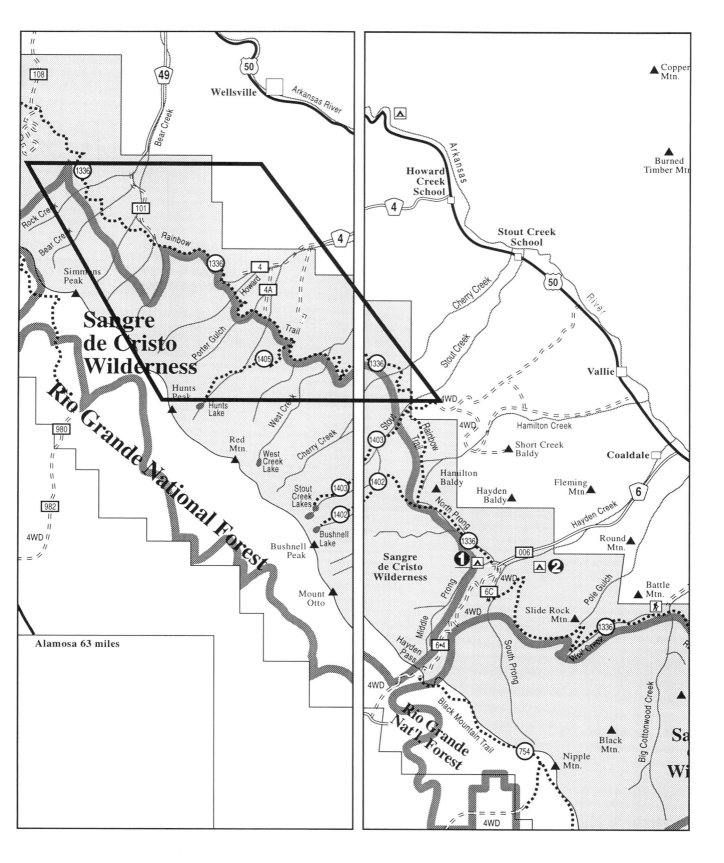

MAP 13 - East Half
TRAIL #1336F-G

MAP 14 - West Half

© *1995 — Outdoor Books & Maps, Inc.* • *Denver, Colorado* • **(303) 629-6111**

ROCKY MOUNTAIN REGION

MAP NUMBER(S): 13

NATIONAL FOREST: SAN ISABEL
RANGER DISTRICT: SALIDA

USGS: WELLSVILLE, PONCHA PASS QUADS

DIFFICULTY: MODERATE
DISTANCE: 12.3 MILES

TRAIL 1336G-H
BEAR CREEK-HIGHWAY 285

UPDATED: OCTOBER 1987

TRAIL BEGINNING ELEVATION:
8,880 ft. Bear Creek, south of Salida.

TRAIL ENDING ELEVATION:
8,540 ft. U.S. Hwy 285, 5 mi. south of Poncha Springs

ACCESS:

Highway 285: The Rainbow Trail crosses US Hwy 285 about 5 miles south of Poncha Springs. Co. Parking space is limited. Please be careful if you are crossng the highway.

Methodist Mtn: At Salida, take County Road 107 south to County Road 108. CR 108 at about 2 miles becomes a very rough Forest Road for 1-1/2 miles to the trail Crossing. Forest Road 108 is recommended for high clearance vehicles only.

Bear Creek: The trail is accessible from the end of Road No. 101. The road is rough and recommended for high clearance vehicles only.

ATTRACTIONS:

Camping is permitted at undeveloped sites along the trail. Water may not be available from Bear Creek to Hwy 285 during dry periods. Scenery is outstanding with excellent views of the Upper Arkansas Valley looking northward to Leadville and Colorado's highest peaks along the Sawatch Range. Methodist Mtn Road No. 108 passes thru private land above the trail and is not open to the public beyond that point. Hikers may by pass the private land by following the powerline and continue to hike up the road. Motorized users may not follow this route however.

This section of the trail is open to motorized trail vehicles. Users are reminded to exercise caution and respect the rights and safety of other users.

USE: Light

RECOMMENDED SEASON:

SPRING | **SUMMER** | **FALL** | WINTER

The trail from Methodist Mountain Road westward is relatively new and does not appear on the USGS or many other maps. Please use the map on the back for reference along with other maps.

NARRATIVE:

TRAIL PROFILE (ONE-WAY):

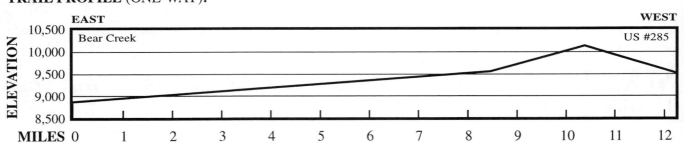

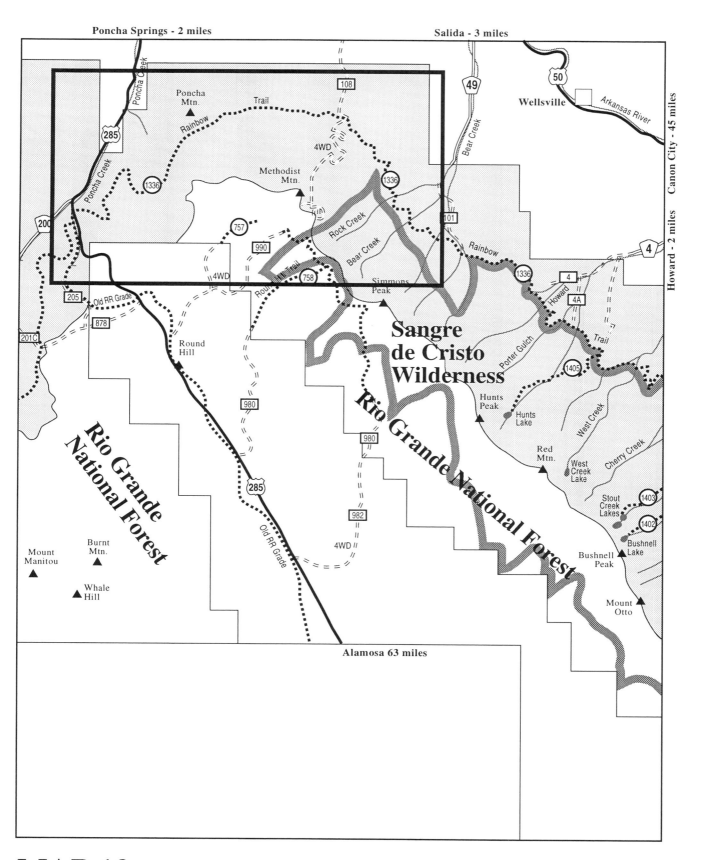

MAP 13
TRAIL #1336G-H

ROCKY MOUNTAIN REGION

MAP NUMBER(S): 12 and 13

NATIONAL FOREST: SAN ISABEL
RANGER DISTRICT: SALIDA

USGS: BONANZA, MOUNT OURAY, PONCHA PASS QUADS

UPDATED: SEPTEMBER 1987

TRAIL BEGINNING ELEVATION:
9,540 ft. U.S. Hwy 285, 5 mi. south of Poncha Springs

TRAIL ENDING ELEVATION:
11,240 ft. Colorado Trail #1776

ACCESS:

Highway 285: The Rainbow Trail crosses US Hwy 285 about 5 miles south of Poncha Springs, Co. Parking space is limited. Please be careful if you are crossing the highway.

Silver Creek: From US 285 about 4.5 miles south of Poncha Springs take County Road 200 for 2 miles to Forest Road #201. Trail joins #201 at about 5 miles. The road past Silver Creek Lakes is suitable for high clearance vehicles only.

Marshall Pass: The trail is accessable from the Colorado Trail No. 1776 about 3 miles south of Marshall Pass.

ATTRACTIONS:

Camping is permitted at undeveloped sites along the trail. Silver Creek offers the best opportunities. Silver Creek is a popular but lightly used fishing stream. The Continental Divide marks the west end of the Rainbow Trail where it joins the Colorado Trail. The Colorado Trail No. 1776 is a 450 mile trail now going from Denver to Durango. Eventually when completed the Continental Divide Trail will pass at this point too. The terrain and life zones on this segment are varied and interesting. The section of the trail from Silver Creek at Toll Road Gulch to the end at the Continental Divide is closed to motor vehicles. Mining improvements at the Kismuth Mine at the area known as The Gate are private Property. Please do not trespass.

DIFFICULTY: MODERATE

DISTANCE: 13.0 MILES

TRAIL 1336H-I

HIGHWAY 285-COLORADO TRAIL #1776

USE: Light

RECOMMENDED SEASON:

| SPRING | **SUMMER** | **FALL** | WINTER |

NARRATIVE:

TRAIL PROFILE (ONE-WAY):

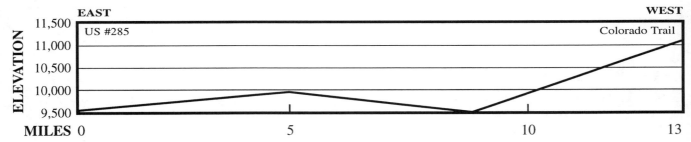

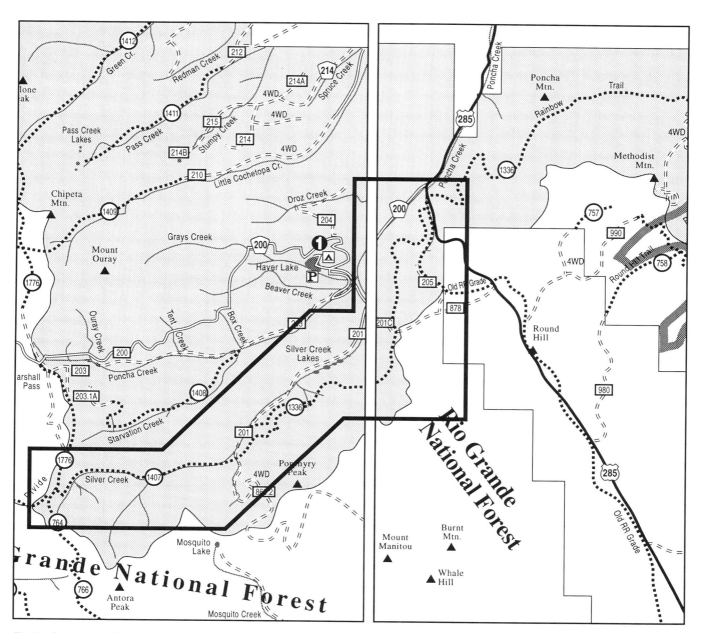

MAP 12 - East Half   **MAP 13 -** West Half

TRAIL #1336H-I

© *1995 — Outdoor Books & Maps, Inc. • Denver, Colorado •* **(303) 629-6111**

ROCKY MOUNTAIN REGION

MAP NUMBER(S): 18

NATIONAL FOREST: SAN ISABEL
RANGER DISTRICT: SAN CARLOS

USGS: CRESTONE PEAK QUAD

UPDATED:

TRAIL BEGINNING ELEVATION:
11,040 ft. South Colony Creek

TRAIL ENDING ELEVATION:
12,030 ft. Upper South Colony Lake

ACCESS:
18 miles southwest of Westcliffe proceed to South Colony Road (FSR #120) which is a 4-Wheel Drive road.

ATTRACTIONS:
Scenery, South Colony Lakes, Crestone Needle, fishing in Sangre De Cristo Wilderness.

DIFFICULTY: DIFFICULT

DISTANCE: 2.0 MILES

TRAIL 1339
SOUTH COLONY

USE: Moderate

RECOMMENDED SEASON: SPRING SUMMER FALL WINTER

NARRATIVE:

TRAIL PROFILE (ONE-WAY):

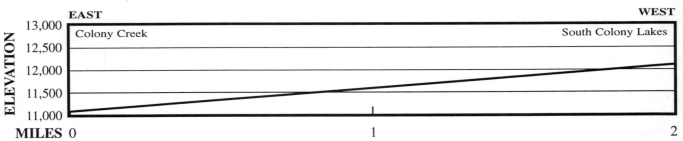

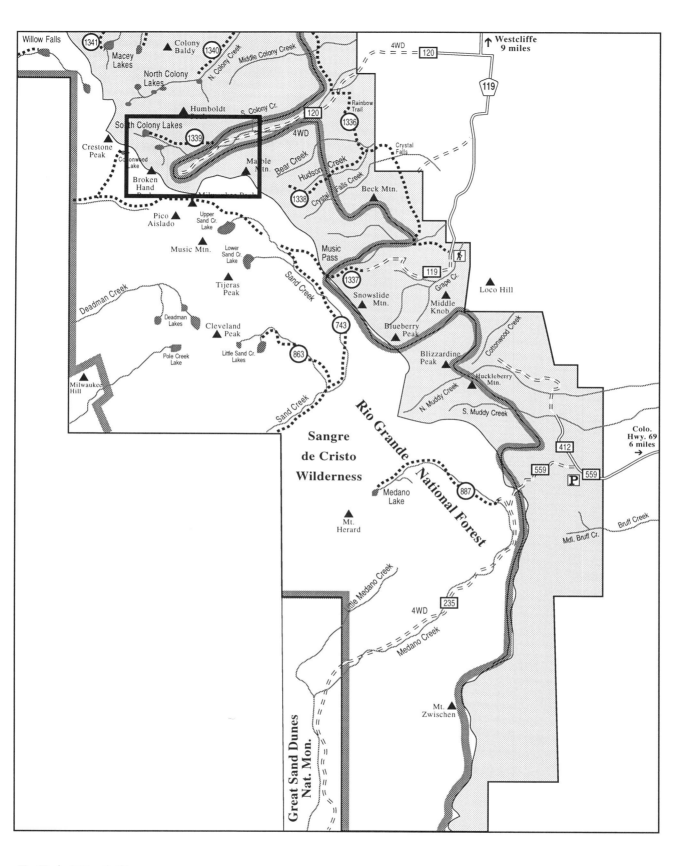

MAP 18
TRAIL #1339

© *1995 — Outdoor Books & Maps, Inc.* • *Denver, Colorado* • (303) 629-6111

ROCKY MOUNTAIN REGION

MAP NUMBER(S): 16 and 18

NATIONAL FOREST: SAN ISABEL
RANGER DISTRICT: SAN CARLOS

USGS: HORN PEAK, CRESTONE PEAK QUADS

DIFFICULTY: DIFFICULT

DISTANCE: 3.0 MILES

TRAIL
1340
NORTH COLONY

UPDATED:

TRAIL BEGINNING ELEVATION:
9,800 ft. Rainbow Trail

TRAIL ENDING ELEVATION:
11,600 ft. North Colony Lakes

ACCESS:
15 miles southwest of Westcliffe proceed to South Colony Road, FSR #313 then to the Rainbow Trail.

ATTRACTIONS:
Scenery, North Colony Lakes, fishing in Sangre De Cristo Wilderness.

USE: Moderate

RECOMMENDED SEASON:

| SPRING | SUMMER | FALL | WINTER |

NARRATIVE:

TRAIL PROFILE (ONE-WAY):

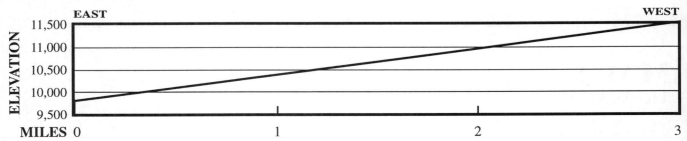

MAP 16 - South East Quarter

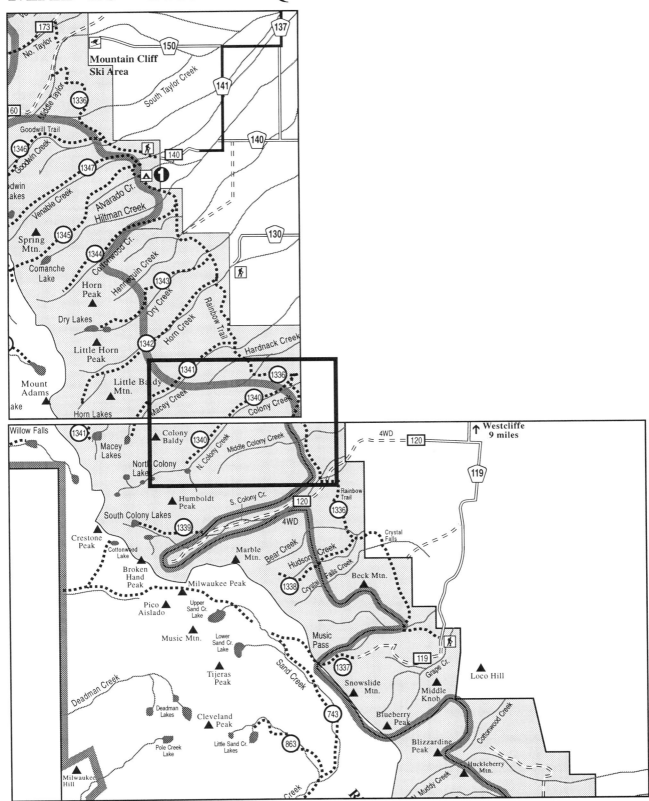

MAP 18 - North West Half
TRAIL #1340

© 1995 — Outdoor Books & Maps, Inc. • Denver, Colorado • (303) 629-6111

ROCKY MOUNTAIN REGION

MAP NUMBER(S): 16 and 18

NATIONAL FOREST: SAN ISABEL
RANGER DISTRICT: SAN CARLOS

USGS: HORN PEAK, CRESTONE PEAK QUADS

DIFFICULTY: DIFFICULT

DISTANCE: 3.0 MILES

TRAIL 1341 MACEY

UPDATED:

TRAIL BEGINNING ELEVATION:
10,150 ft. Rainbow Trail

TRAIL ENDING ELEVATION:
11,860 ft. Macey Lakes

ACCESS:
15 miles southwest of Westcliffe proceed to Horn Road to the Rainbow Trail.

ATTRACTIONS:
Scenery, Macey Lakes, fishing, in Sangre De Cristo Wilderness.

USE: Moderate

RECOMMENDED SEASON:
SPRING | **SUMMER** | **FALL** | WINTER

NARRATIVE:

TRAIL PROFILE (ONE-WAY):

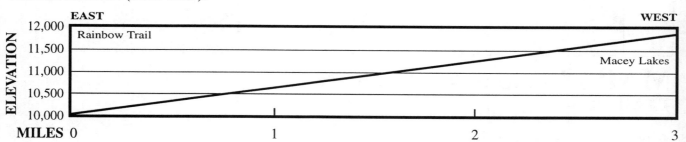

MAP 16 - South East Quarter

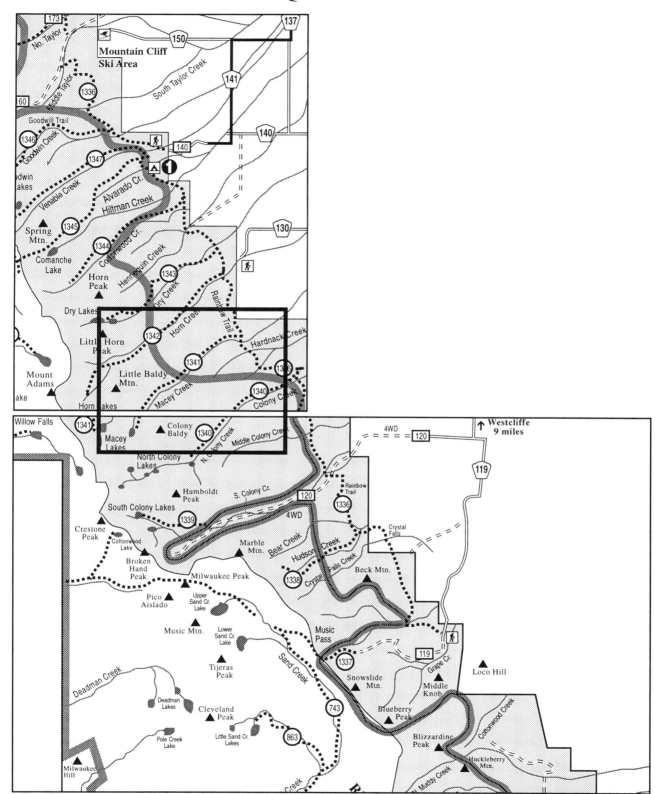

MAP 18 - North West Half
TRAIL #1341

© 1995 — *Outdoor Books & Maps, Inc.* • *Denver, Colorado* • **(303) 629-6111**

ROCKY MOUNTAIN REGION

MAP NUMBER(S): 16

NATIONAL FOREST: SAN ISABEL
RANGER DISTRICT: SAN CARLOS

USGS: HORN PEAK QUAD

DIFFICULTY: DIFFICULT

DISTANCE: 5.0 MILES

TRAIL 1342 HORN CREEK

UPDATED:

TRAIL BEGINNING ELEVATION:
9,320 ft. Rainbow Trail

TRAIL ENDING ELEVATION:
11,840 ft. Horn Lakes

ACCESS:
15 miles southwest of Westcliffe to Horn Road.

ATTRACTIONS:
Scenery, Horn Lakes, fishing, partly in Sangre De Cristo Wilderness.

USE: Moderate

RECOMMENDED SEASON:
SPRING | SUMMER | FALL | WINTER

NARRATIVE:

TRAIL PROFILE (ONE-WAY):

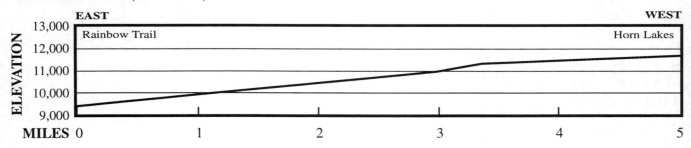

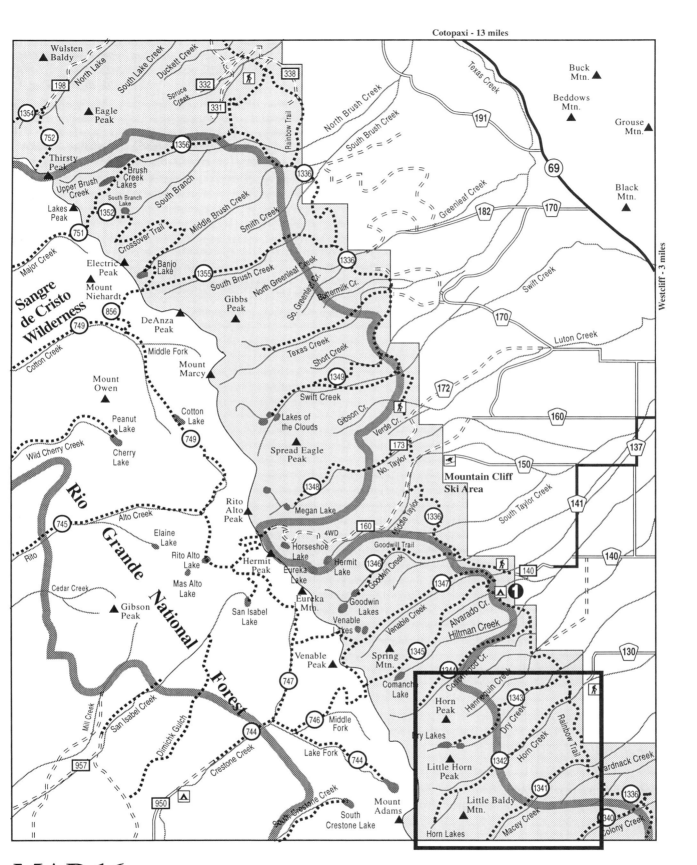

MAP 16
TRAIL #1342

© 1995 — *Outdoor Books & Maps, Inc.* • *Denver, Colorado* • **(303) 629-6111**

ROCKY MOUNTAIN REGION

MAP NUMBER(S): 16

NATIONAL FOREST: SAN ISABEL
RANGER DISTRICT: SAN CARLOS

USGS: HORN PEAK QUAD

DIFFICULTY: DIFFICULT

DISTANCE: 3.0 MILES

TRAIL 1343
DRY CREEK

UPDATED:

TRAIL BEGINNING ELEVATION:
9,200 ft. Rainbow Trail

TRAIL ENDING ELEVATION:
11,840 ft. Dry Lakes

ACCESS:
15 miles southwest of Westcliffe to Horn Road then proceed to Rainbow Trail.

ATTRACTIONS:
Dry Lakes, scenery, fishing, partly in Sangre De Critsto Wilderness.

USE: Moderate

RECOMMENDED SEASON:
SPRING | **SUMMER** | **FALL** | WINTER

NARRATIVE:

TRAIL PROFILE (ONE-WAY):

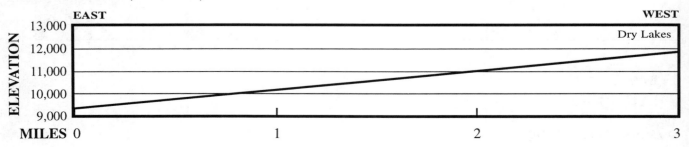

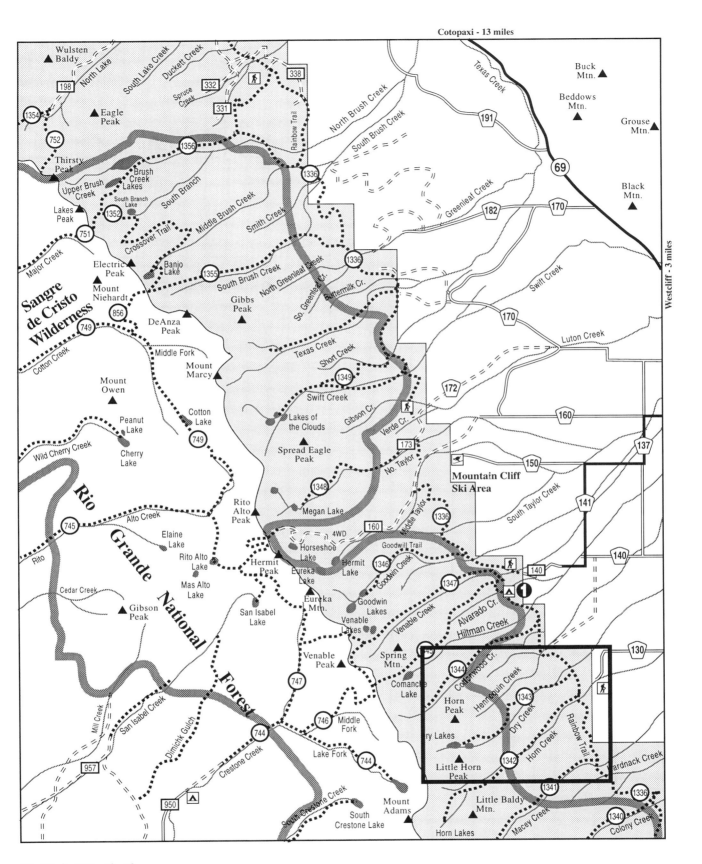

MAP 16
TRAIL #1342

© 1995 — Outdoor Books & Maps, Inc. • Denver, Colorado • (303) 629-6111

ROCKY MOUNTAIN REGION

MAP NUMBER(S): 16

NATIONAL FOREST: SAN ISABEL
RANGER DISTRICT: SAN CARLOS

USGS: HORN PEAK QUAD

UPDATED:

TRAIL BEGINNING ELEVATION:
9,400 ft. Rainbow Trail

TRAIL ENDING ELEVATION:
11,600 ft. Cottonwood Creek Basin

ACCESS:
10 miles southwest of Westcliffe, Forest Service Road #302, Rainbow Trail.

ATTRACTIONS:
Scenery. No facilities.

DIFFICULTY: MORE DIFFICULT
DISTANCE: 4.0 MILES

TRAIL 1344
COTTONWOOD

USE: Moderate

RECOMMENDED SEASON:

| SPRING | SUMMER | FALL | WINTER |

NARRATIVE:

TRAIL PROFILE (ONE-WAY):

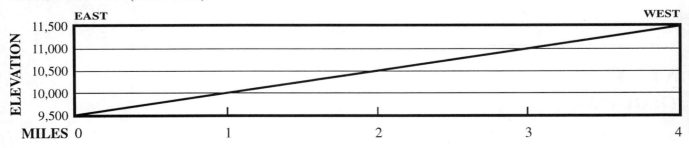

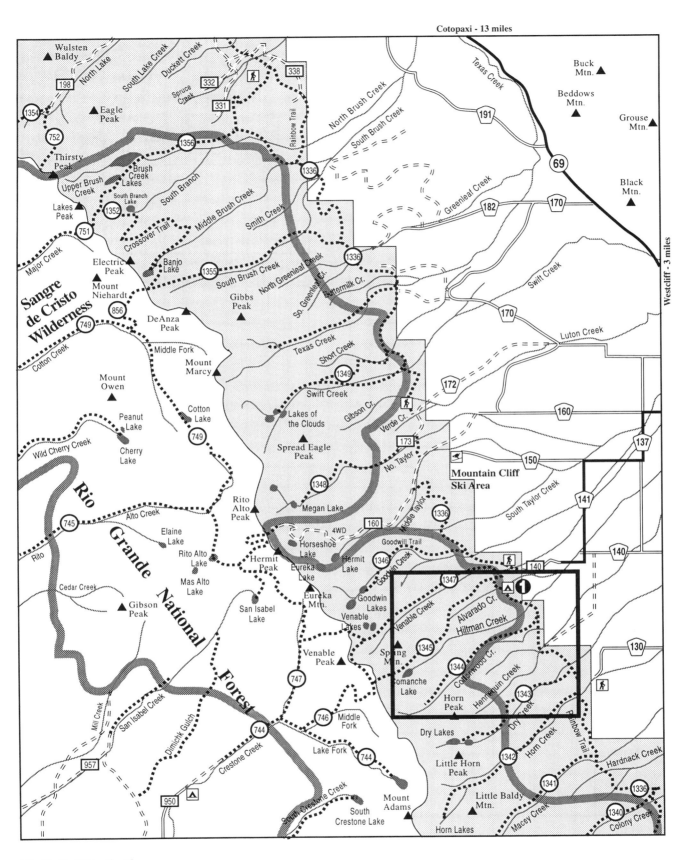

MAP 16
TRAIL #1344

© 1995 — Outdoor Books & Maps, Inc. • Denver, Colorado • (303) 629-6111

ROCKY MOUNTAIN REGION

MAP NUMBER(S): 16

NATIONAL FOREST: SAN ISABEL
RANGER DISTRICT: SAN CARLOS

USGS: HORN PEAK, RITO ALTO PEAK QUADS

UPDATED: MARCH 1988

TRAIL BEGINNING ELEVATION:
9,000 ft., Alvarado Campground

TRAIL ENDING ELEVATION:
12,800 ft. on Rainbow Trail, 5 mi. north of Alvarado Campground

ACCESS:
The Alvarado Campground is located about nine miles southwest of Westcliffe, Colorado on County Roads and Forest Road #302. The campground is heavily used during the summer and the trailhead facilities serve as the focal point for a network of trails. Persons using the trails only should park in the trailhead parking lots rather than the campground units. The trailhead is suitable for moderate sized trucks and horse or trail vehicle trailers.

ATTRACTIONS:
The Comanche-Venable Trail makes a loop from Alvarado Campground past Comanche Lake to the crest of the Sangre de Cristo Range, across Phantom-Terrace and back down Venable Creek past Venable Lakes to the Rainbow Trail a half-mile north of the starting point. This is a very popular and heavily used trail throughout the summer.

For hikers, the trail is moderate to more difficult because of sustained grades and steep pitches. Although the average grade is 14% on the south section and 17% on the north, some short sections reach 23%. The trail is in good condition and offers outstanding scenery and views of the valley below. Hikers should plan a full day to make the round trip. Many hikers continue the hike over the divide to the west side of the Sangre on the Rio Grande National Forest for extended trips.

Horse use is common on the trail, especially for extended trips. Special care should be taken. There is little forage available and overnight users should pack adequate feed.

TRAIL PROFILE (ONE-WAY):

TRAIL 1345(1347)
COMANCHE-VENABLE LOOP

DIFFICULTY: DIFFICULT
DISTANCE: 10.5 MILES

USE: Very Heavy

RECOMMENDED SEASON:

SPRING	SUMMER	FALL	WINTER

Phantom-Terrace and other sections are narrow with very steep sideslopes and the rider should use caution.

The area receives heavy overnight use particularly around the lakes. Your help is needed to keep the area clean and attractive. Please pack out your trash and any other you may find. Avoid camping on the lakeshore or stream banks. Be especially careful with fire. Small camp stoves are recommended. Horses should be tied away from waters' edge and other campsites.

NARRATIVE:
Trail CLOSED to motorized vehicles: #1342 Horn Creek, #1343 Dry Creek, #1344 Cottonwood, #1345 Comanche-Venable Phantom-Terrace #1346 Goodwin, trails in Rio Grande National Forest are: #744, 745, 746, 747, 859 and 1345

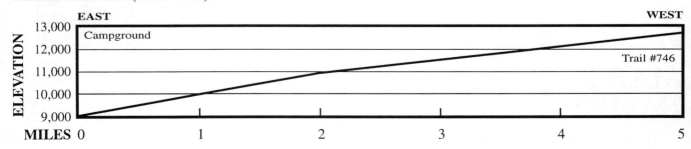

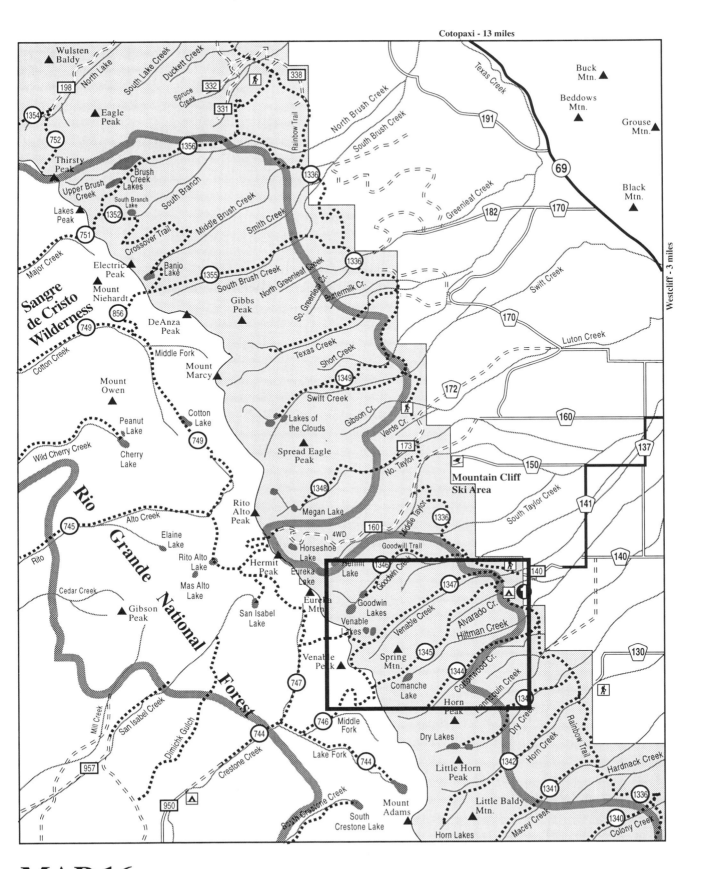

MAP 16
TRAIL #1345(1347)

© *1995 — Outdoor Books & Maps, Inc.* • *Denver, Colorado* • **(303) 629-6111**

ROCKY MOUNTAIN REGION

MAP NUMBER(S): 16

NATIONAL FOREST: SAN ISABEL
RANGER DISTRICT: SAN CARLOS

USGS: HORN PEAK, QUAD

DIFFICULTY: DIFFICULT

DISTANCE: 4.0 MILES

TRAIL 1346
GOODWIN LAKE

UPDATED:

TRAIL BEGINNING ELEVATION:
9,600 ft. Ranbow Trail

TRAIL ENDING ELEVATION:
11,450 ft. Goodwin Lakes

ACCESS:
10 miles southwest of Westcliffe to FSR #302.

ATTRACTIONS:
Scenery, Goodwin Lakes in Sangre De Cristo Wilderness.

USE: Moderate

RECOMMENDED SEASON:

| SPRING | SUMMER | FALL | WINTER |

NARRATIVE:

TRAIL PROFILE (ONE-WAY):

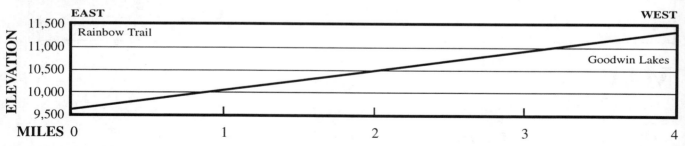

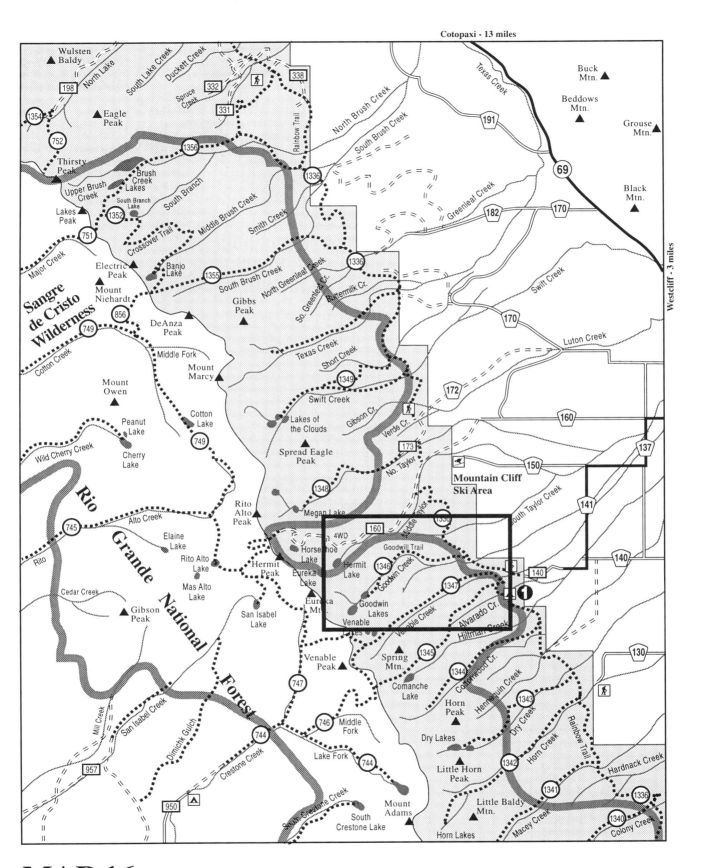

MAP 16
TRAIL #1346

© 1995 — Outdoor Books & Maps, Inc. • Denver, Colorado • (303) 629-6111

ROCKY MOUNTAIN REGION
MAP NUMBER(S): 16

NATIONAL FOREST: SAN ISABEL
RANGER DISTRICT: SAN CARLOS

USGS: ELECTRIC PEAK, BECKWITH MTN. QUADS

DIFFICULTY: DIFFICULT

DISTANCE: 6.5 MILES

TRAIL 1349
LAKE OF THE CLOUDS

UPDATED: AUGUST 1984

TRAIL BEGINNING ELEVATION:
9,200 ft. Rainbow Trail

TRAIL ENDING ELEVATION:
11,600 ft. Lakes of the Clouds

USE: Moderate

RECOMMENDED SEASON:
SPRING | **SUMMER** | **FALL** | WINTER

ACCESS:
Drive up Hermit road to where the jeep road intersects the Rainbow Trail. The sign is hard to find. It is off to the right on the north side of the road, after you get into the trees. There is no designated parking lot, but there is ample parking spaces along this road. You will have to walk north on the Rainbow Trail for 2 miles (gentle walking) before you will reach the Lake of the Clouds trailhead, it is marked. Then it is approximately 4.5 miles to the lake. This part of the trail is steep and rocky, but there is no danger involved.

ATTRACTIONS:
Scenery, Lakes of the Clouds, in the Sangre De Cristo Wilderness.

NARRATIVE:

TRAIL PROFILE (ONE-WAY):

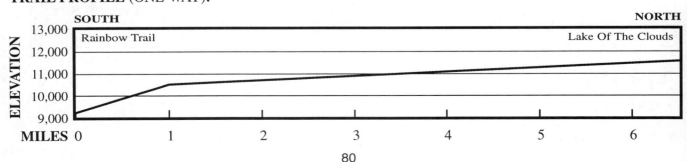

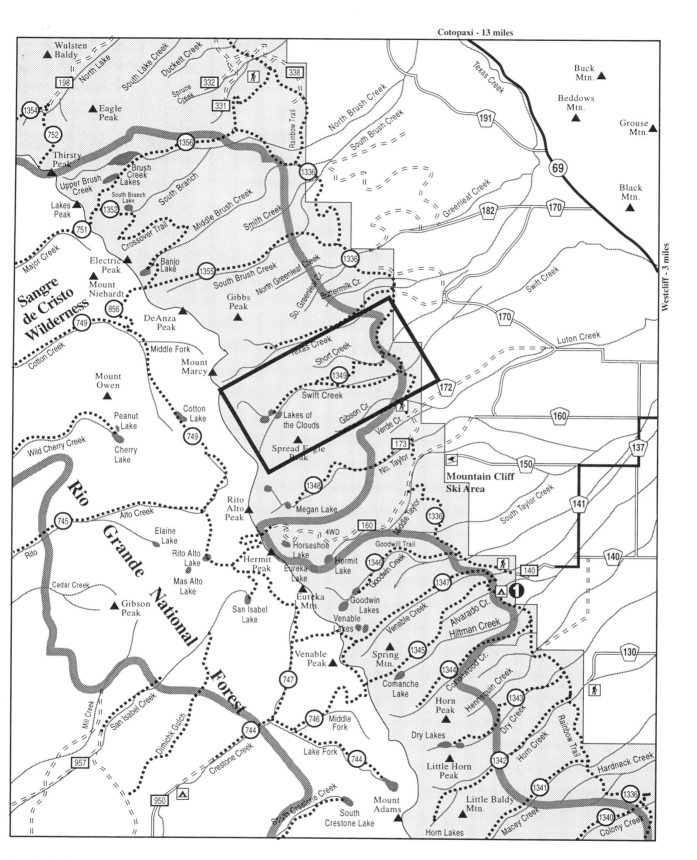

MAP 16
TRAIL #1349

ROCKY MOUNTAIN REGION

MAP NUMBER(S): 16

NATIONAL FOREST: SAN ISABEL
RANGER DISTRICT: SAN CARLOS

USGS: ELECTRIC PEAK QUAD

DIFFICULTY: DIFFICULT

DISTANCE: 6.0 MILES

TRAIL 1355
SOUTH BRUSH CREEK

UPDATED: FEBRUARY 1987

TRAIL BEGINNING ELEVATION:
9,300 ft. Rainbow Trail

TRAIL ENDING ELEVATION:
12,820 ft. Forest Boundary

USE: Moderate

RECOMMENDED SEASON:

| SPRING | SUMMER | FALL | WINTER |

ACCESS:
Access via Hwy 69 to Hillside, west on Forest Road #300 for 2 miles. Then turn left onto Forest Road #332, which is a rough rocky dirt road. Follow this road down across Spruce Creek. Turn right onto Forest Road #337 and follow to Brush Lakes Trailhead. Follow Rainbow Trail about 5 miles south to junction of South Brush Creek Trail. Access can also be obtained from the South via Gibson Creek Trailhead.

ATTRACTIONS:
Scenery, Banjo Lake, fishing, last 5 miles in Sangre De Cristo Wilderness.

NARRATIVE:

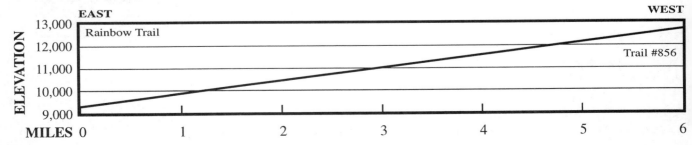

TRAIL PROFILE (ONE-WAY):

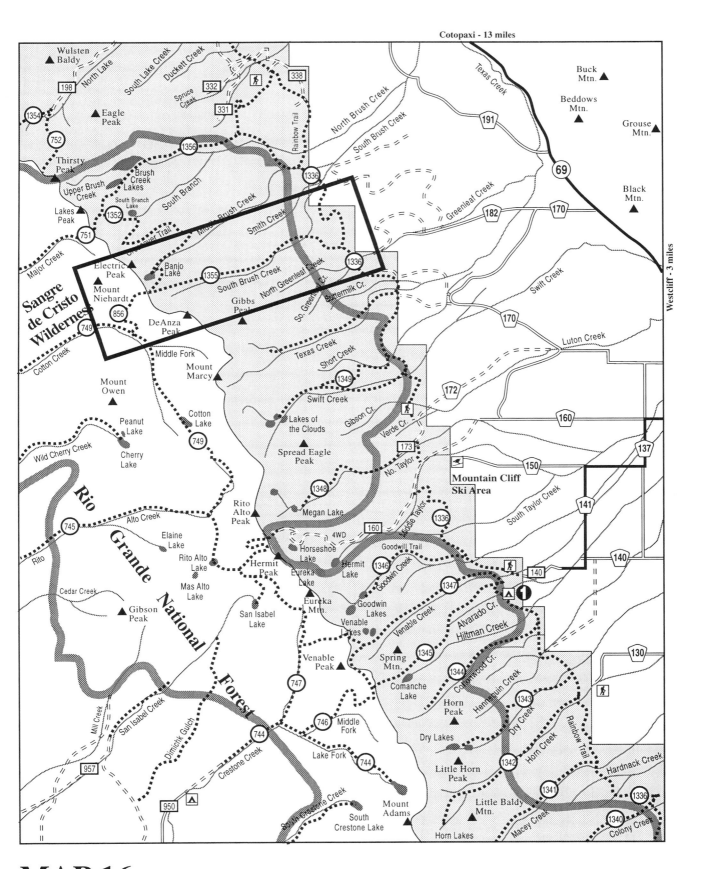

MAP 16
TRAIL #1355

© 1995 — *Outdoor Books & Maps, Inc.* • *Denver, Colorado* • **(303) 629-6111**

ROCKY MOUNTAIN REGION

MAP NUMBER(S): 16

NATIONAL FOREST: SAN ISABEL
RANGER DISTRICT: SAN CARLOS

USGS: ELECTRIC PEAK QUAD

DIFFICULTY: DIFFICULT

DISTANCE: 7.0 MILES

TRAIL 1356

NORTH BRUSH CREEK

UPDATED:

TRAIL BEGINNING ELEVATION:
8,540 ft. Brush Creek Trailhead

TRAIL ENDING ELEVATION:
12,480 ft. Forest Boundary

ACCESS:
5 miles west of Hillside on Hwy 69 to FDR #332 to FDR #337. Then follow signs to North Brush Creek access trailhead.

ATTRACTIONS:
Scenery, Brush Creek Lakes, Fishing, Deer, Elk, Rocky Mtn. Big Horn sheep, Blue Grouse within Sangre De Cristo Wilderness.

USE: Moderate

RECOMMENDED SEASON:
SPRING | **SUMMER** | **FALL** | WINTER

NARRATIVE:

TRAIL PROFILE (ONE-WAY):

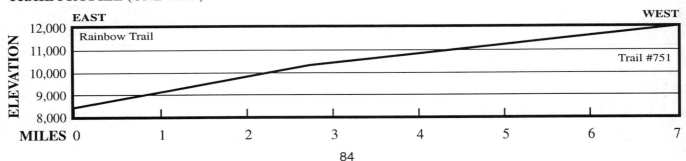

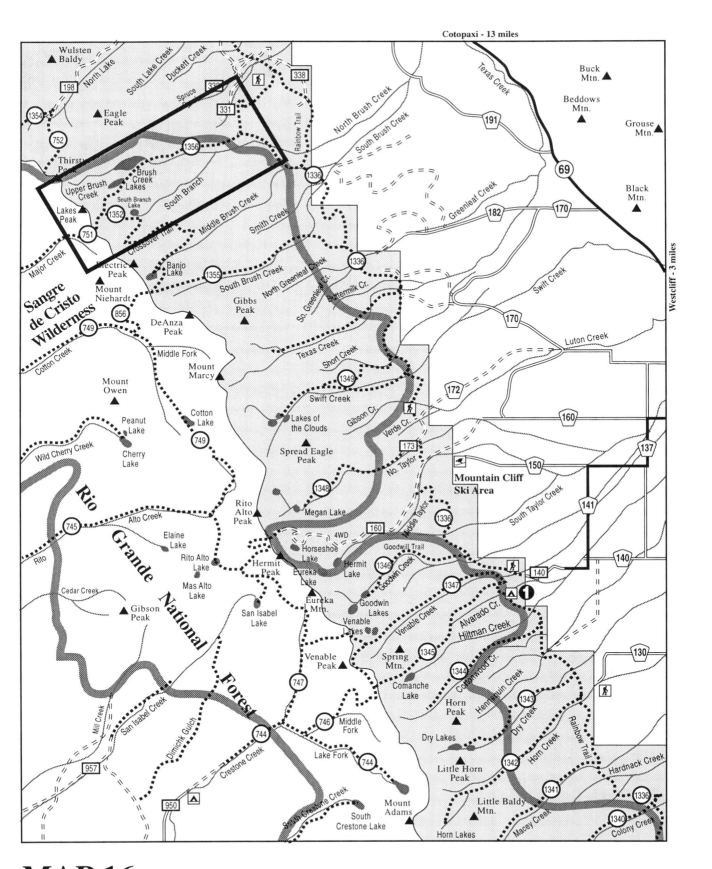

MAP 16
TRAIL #1356

© *1995 — Outdoor Books & Maps, Inc.* • *Denver, Colorado* • **(303) 629-6111**

ROCKY MOUNTAIN REGION
MAP NUMBER(S): 6

NATIONAL FOREST: SAN ISABEL & PIKE
RANGER DISTRICT: LEADVILLE

USGS: MT. HARVARD, HARVARD LAKES QUADS

UPDATED: MARCH 1988

TRAIL BEGINNING ELEVATION:
9,100 ft. Intersection of Trail #1776

TRAIL ENDING ELEVATION:
12,600 ft. Intersection of Trail #1469

ACCESS:
Two (2) miles south of Clear Creek Reservoir or 12 miles north of Buena Vista on U.S. Highway 24, turn west on Forest Road 123. Travel about two (2) miles west on this single lane gravel road to the beginning of the trail. Vehicles with low clearance should be parked near the highway.

Drinking water, toilets, or overnight facilities are not available at the trailhead.

Some visitors come into the area via the Missouri Gulch Trail from Clear Creek or by the Colorado Trail from the south.

ATTRACTIONS:
This trail is for foot and horse use only. Motorized vehicles are not permitted. Hikers and horseback riders are requested to stay on constructed trails wherever possible.

DIFFICULTY: MODERATE

DISTANCE: 10.0 MILES

TRAIL 1374
PINE CREEK

USE: Heavy (Approx. 150 people per day)
No motorized vehicles

RECOMMENDED SEASON:

| SPRING | SUMMER | FALL | WINTER |

NARRATIVE:

TRAIL PROFILE (ONE-WAY):

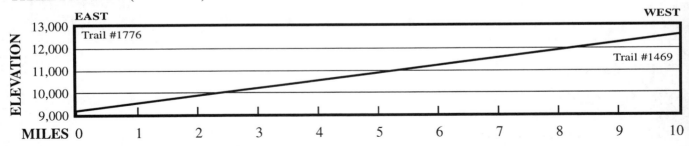

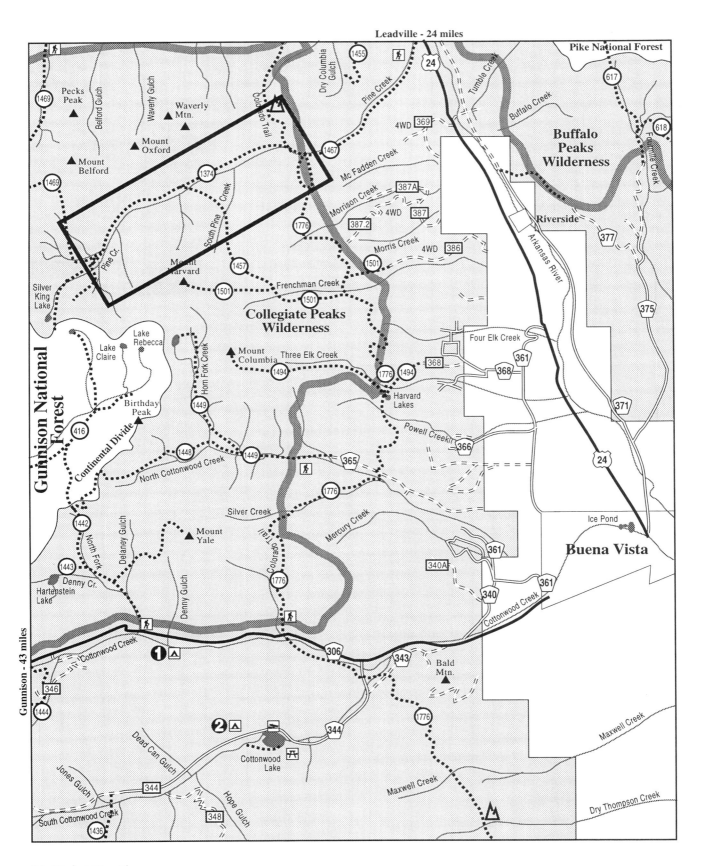

MAP 6
TRAIL #1374

© *1995 — Outdoor Books & Maps, Inc.* • *Denver, Colorado* • (303) 629-6111

ROCKY MOUNTAIN REGION
MAP NUMBER(S): 5 and 6

NATIONAL FOREST: SAN ISABEL
RANGER DISTRICT: LEADVILLE

USGS: MT. HARVARD, WINFIELD QUADS

DIFFICULTY: MORE TO VERY DIFFICULT
DISTANCE: 4.0 MILES

TRAIL 1378
MISSOURI MOUNTAIN

UPDATED: DECEMBER 1990

USE: Moderate to Heavy

TRAIL BEGINNING ELEVATION:
9,500 ft.

TRAIL ENDING ELEVATION:
14,067 ft Summit of Missouri Mountain

RECOMMENDED SEASON:

SPRING	SUMMER	FALL	WINTER

ACCESS:
#1 From Leadville travel south on US Hwy 24 to Granite, continue 1-1/2 miles S to Clear Creek Road, Forest Road 390, travel west 10 miles to Rockdale. The trailhead provides parking space but has no restrooms, water or overnight facilities.

NARRATIVE:
Rockdale, Lake Fork of Clear Cr. 4WD can continue 3 miles to road closure at Cloyses Lake, otherwise walk, path forks East at road closure, climbs steeply to glacial basin approximately 12,000' elevation. Fork northeast, climb to summit ridge, attain ridge at approximately 13,000' elevation. hike along ridge to summit, 14,067 feet

ACCESS:
#2: From Leadville travel south on US Hwy 24 to Granite, continue 1- 1/2 miles south to Clear Creek Road, Forest Road 390 8 miles to Vicksburg, park in Missouri Gulch parking. (Length 2-5 miles)

NARRATIVE:
From Missouri Gulch Trailhead lot near Vicksburg. Hike up trail to glacial basin above treeline, at approximate elevation of 12,600', fork southwest from trail across basin to summit ridge west 1/2 mile from peak. Hike ridge east to peak.

ACCESS:
#3: From Leadville travel south on US Hwy 24 to Gold Camp Trailer Park. Continue up dirt road to road closure/trailhead.(Forest Road 388)

NARRATIVE:
Pine Creek trailhead three miles W of US Hwy 24 on Pine Creek Trail, past Colorado Trail, past Little John's Cabins to Missouri Basin Trail to Missouri Basin Fork west off trail to saddle between Iowa Peak and Missouri Mtn. Climb ridge north to summit, 14,067'. (Length 3 -11.5 miles)

NOTE: No well defined or marked trail exists on all three routes above treeline to summit.

TRAIL PROFILE (ONE-WAY):

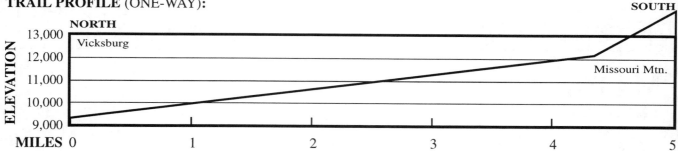

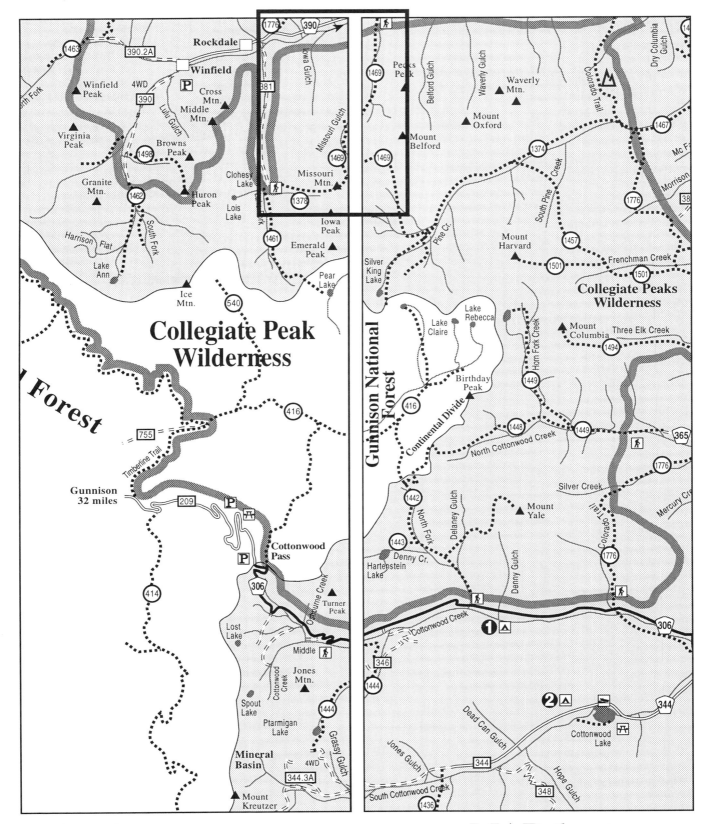

MAP 5 - West Half

MAP 6 - East Half

TRAIL #1378

© 1995 — Outdoor Books & Maps, Inc. • Denver, Colorado • (303) 629-6111

ROCKY MOUNTAIN REGION

MAP NUMBER(S): 17

NATIONAL FOREST: SAN ISABEL
RANGER DISTRICT: SAN CARLOS

USGS: ST. CHARLES PEAK QUAD

DIFFICULTY: MODERATE

DISTANCE: 7.0 MILES

TRAIL 1384
SQUIRREL CREEK

UPDATED: DECEMBER 1987

TRAIL BEGINNING ELEVATION:
6,700 ft. Pueblo Mtn. Park

TRAIL ENDING ELEVATION:
8,600 ft. Campground

ACCESS:
.5 Mile south of Valley View on Colo. #78, west to Pueblo Mtn. Park. West 7 miles to Davenport Campground.

ATTRACTIONS:
Scenery, fishing.

Facilities: Davenport Campground.

USE: Moderate to Heavy
NOTE: This trail is difficult for trailbikes.

RECOMMENDED SEASON:

SPRING	SUMMER	FALL	WINTER

NARRATIVE:

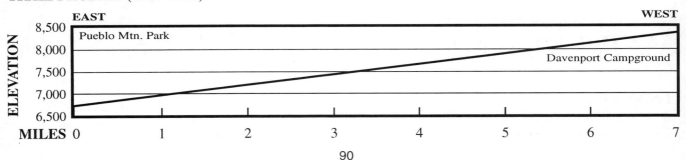

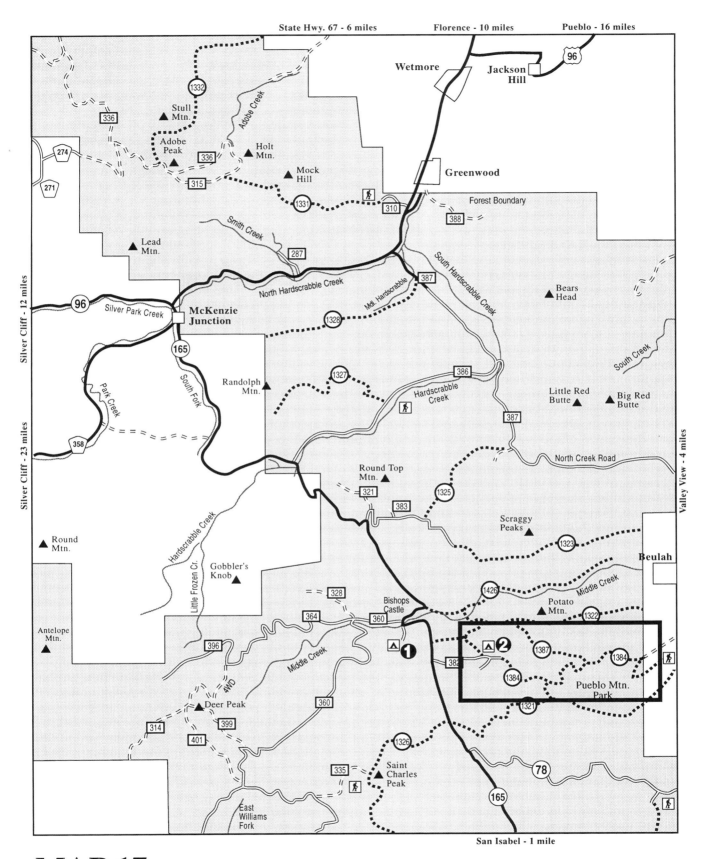

MAP 17
TRAIL #1384

© 1995 — Outdoor Books & Maps, Inc. • Denver, Colorado • (303) 629-6111

ROCKY MOUNTAIN REGION

MAP NUMBER(S): 17

NATIONAL FOREST: SAN ISABEL
RANGER DISTRICT: SAN CARLOS

USGS: ST. CHARLES PEAK QUAD

DIFFICULTY: DIFFICULT

DISTANCE: 2.0 MILES

TRAIL 1387
DOME ROCK

UPDATED:

TRAIL BEGINNING ELEVATION:
8,600 ft. Second Mace Trail

TRAIL ENDING ELEVATION:
7,280 ft. Squirrel Creek

ACCESS:
2 miles west of Beulah, Squirrel Creek Trail & Second Mace Trail. Connects Trails #1322 and #1384.

ATTRACTIONS:
Scenery.

USE: Moderate

RECOMMENDED SEASON:

| SPRING | SUMMER | FALL | WINTER |

NARRATIVE:

TRAIL PROFILE (ONE-WAY):

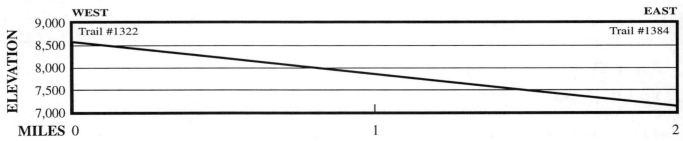

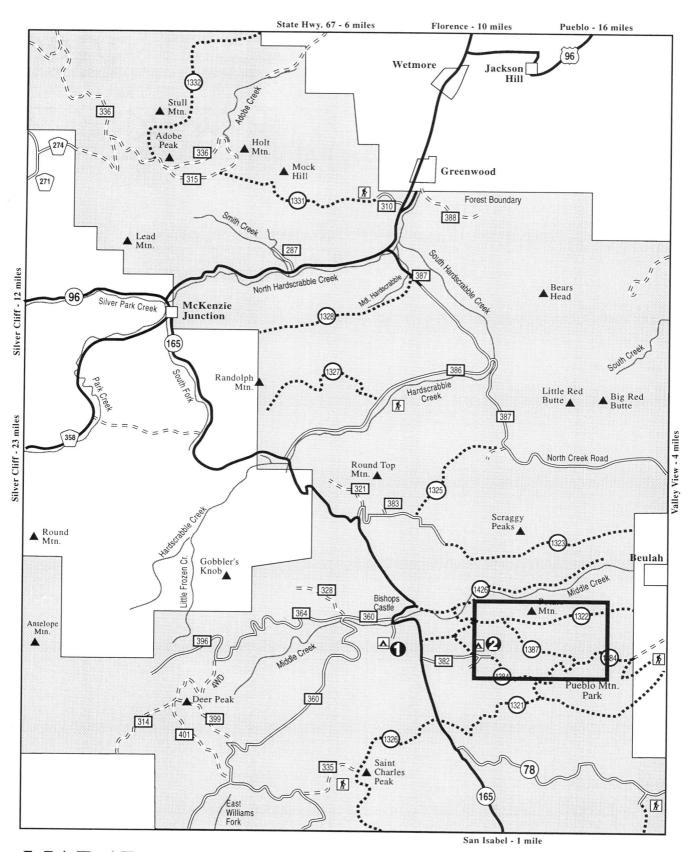

MAP 17
TRAIL #1387

© *1995 — Outdoor Books & Maps, Inc.* • *Denver, Colorado* • **(303) 629-6111**

ROCKY MOUNTAIN REGION

MAP NUMBER(S): 22

NATIONAL FOREST: SAN ISABEL
RANGER DISTRICT:

USGS: CHUCHARAS PASS QUAD

UPDATED:

TRAIL BEGINNING ELEVATION:
11,250 ft. Cordova Pass

TRAIL ENDING ELEVATION:
11,840 ft. West Spanish Peak

ACCESS:
20 miles south of La Veta, Highway 12, Forest Service Road #415.

ATTRACTIONS:
Scenery.

DIFFICULTY: EASY

DISTANCE: 2.5 MILES

TRAIL 1390
WEST SPANISH PEAK

USE:

RECOMMENDED SEASON: SPRING | SUMMER | FALL | WINTER

NARRATIVE:

TRAIL PROFILE (ONE-WAY):

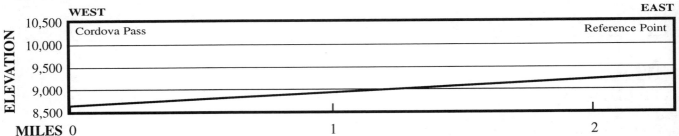

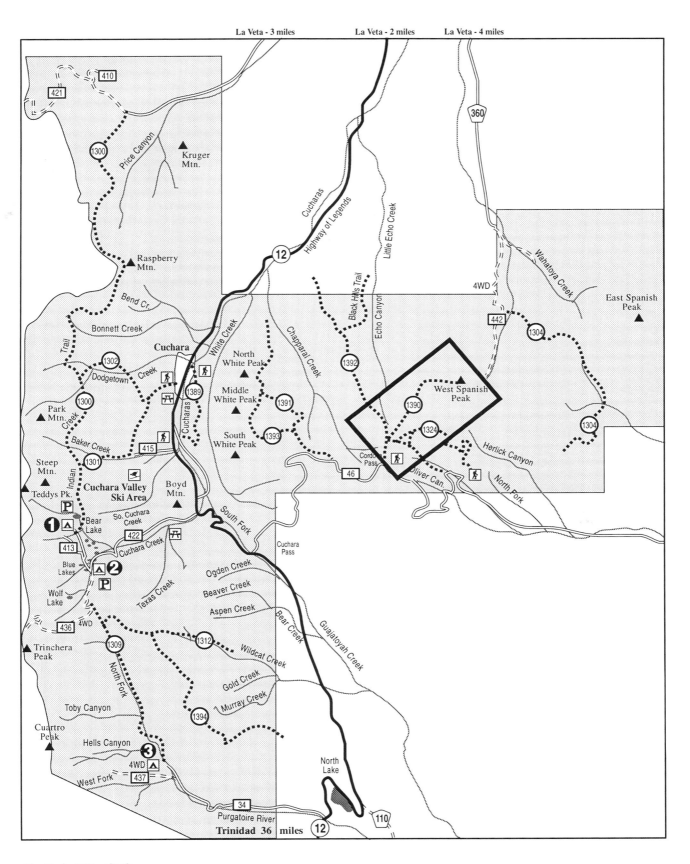

MAP 22
TRAIL #1390

© 1995 — Outdoor Books & Maps, Inc. • Denver, Colorado • (303) 629-6111

ROCKY MOUNTAIN REGION
MAP NUMBER(S): 9

NATIONAL FOREST: SAN ISABEL & PIKE
RANGER DISTRICT: SALIDA

USGS: MOUNT ANTERO QUAD

DIFFICULTY: EASY

DISTANCE: 1.5 MILES

TRAIL 1427
WAGON ROAD LOOP

UPDATED: MARCH 1988

TRAIL BEGINNING ELEVATION:
8,970 ft. FSR #272

TRAIL ENDING ELEVATION:
9,600 ft. Intersection Trail #1776

USE: Light
No motorized vehicles

RECOMMENDED SEASON:
SPRING | SUMMER | FALL | WINTER

ACCESS:
From U.S. 285, midway between Poncha Springs and Buena Vista, take County Road No. 270 west for 1 1/2 miles to County Road 272. Follow 272 straight ahead for 2 miles to intersection, continue to left for another 1 1/2 miles to Browns Creek Trailhead.

ATTRACTIONS AND CONSIDERATIONS:
This trail provides an opportunity for a relatively easy 1/2 day loop hike or ride out of the Browns Creek Trailhead. Visitors can travel up this trail to the Colorado trail and then return by way of Trail No. 1429.
The scenery is varied and interesting. Both Browns Creek and Little Browns Creek offer trout fishing.

A small holding corral is available at the trailhead. Camping at undeveloped sites is possible at the trailhead, or at a few sites along Browns Creek near the Colorado Trail. Campfire permits are not required but build it only in a safe place and make sure you extinguish it completely.

This trail is for foot and horse only. Motor veichles are prohibited.

NARRATIVE:
Trail begins at Browns Creek Trailhead and ends at Colorado Trail 1776.

TRAIL PROFILE (ONE-WAY):

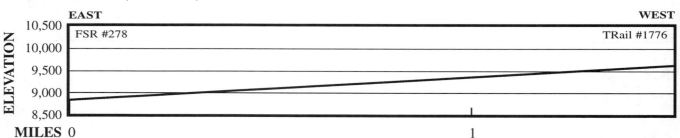

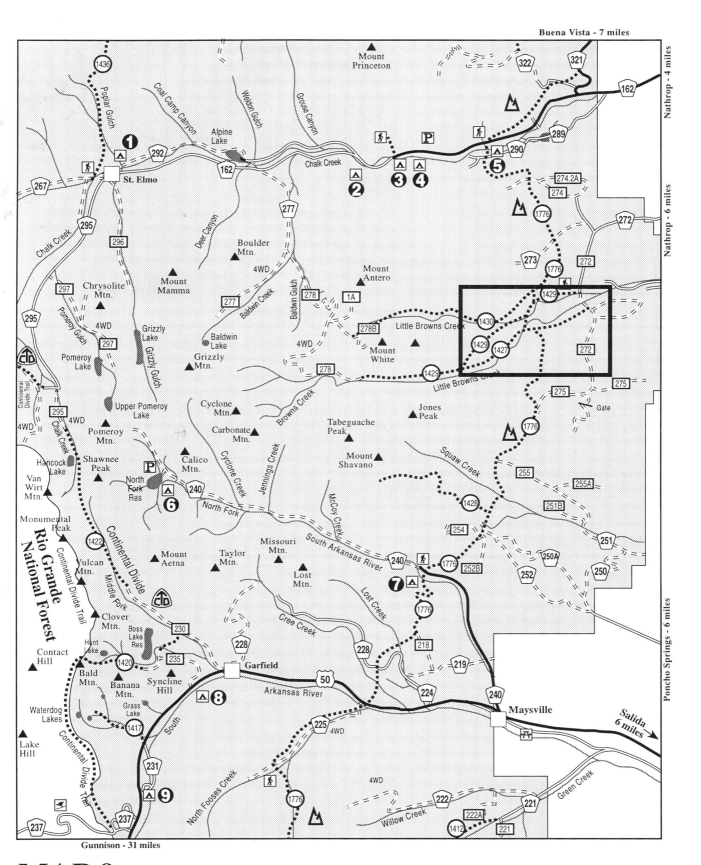

MAP 9
TRAIL #1427

© 1995 — Outdoor Books & Maps, Inc. • Denver, Colorado • (303) 629-6111

ROCKY MOUNTAIN REGION
MAP NUMBER(S): 9

NATIONAL FOREST: SAN ISABEL & PIKE
RANGER DISTRICT: SALIDA

USGS: MOUNT ANTERO QUAD

DIFFICULTY: EASY
DISTANCE: 5.0 MILES

TRAIL 1429
BROWNS CREEK

UPDATED: DECEMBER 1987

TRAIL BEGINNING ELEVATION:
8,970 ft. FSR #272

TRAIL ENDING ELEVATION:
11,600 ft. FSR #278

USE: Moderate
No motorized vehicles

RECOMMENDED SEASON:

| SPRING | SUMMER | FALL | WINTER |

ACCESS:
From U.S. 285, midway between Poncha Springs and Buena Vista, take County Road No. 270 west for 1-1/2 miles to County Road 272. Follow 272 straight ahead for 2 miles to intersection. Continue to left for another 1-1/2 miles to Browns Creek Trailhead.

ATTRACTIONS AND CONSIDERATIONS:
This trail offers outstanding scenery varing from open ponderosa pine, mountain meadows, lodgepole pine and spruce forest to alpine vistas at Mt Antero. Mt Antero at 14,269 feet is one of the major attractions. Loop trips are possible on connecting trails from trailhead. Trout fishng in Browns Creek and Little Browns Creek.

A horse holding corral is available at the trailhead. Camping facilities are not provided. Camping is recommended at undeveloped sites at the trailhead, at the meadow, at Browns Creek (1-1/4 mi.) or at the lake (6 mi.) on Browns Creek. Campfire permits are not required. Please leave your campsite clean and your fire dead out.

This trail is for horses and hikers only. Motorcycles are not allowed.

Four-wheel drive roads No. 277 and 278 provide access to Mt. Antero from Chalk Creek late July through early fall.

NARRATIVE:

TRAIL PROFILE (ONE-WAY):

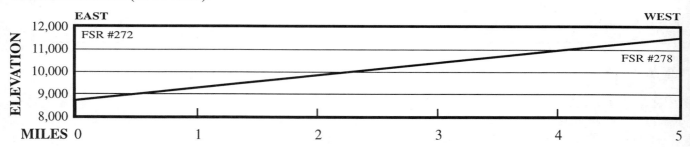

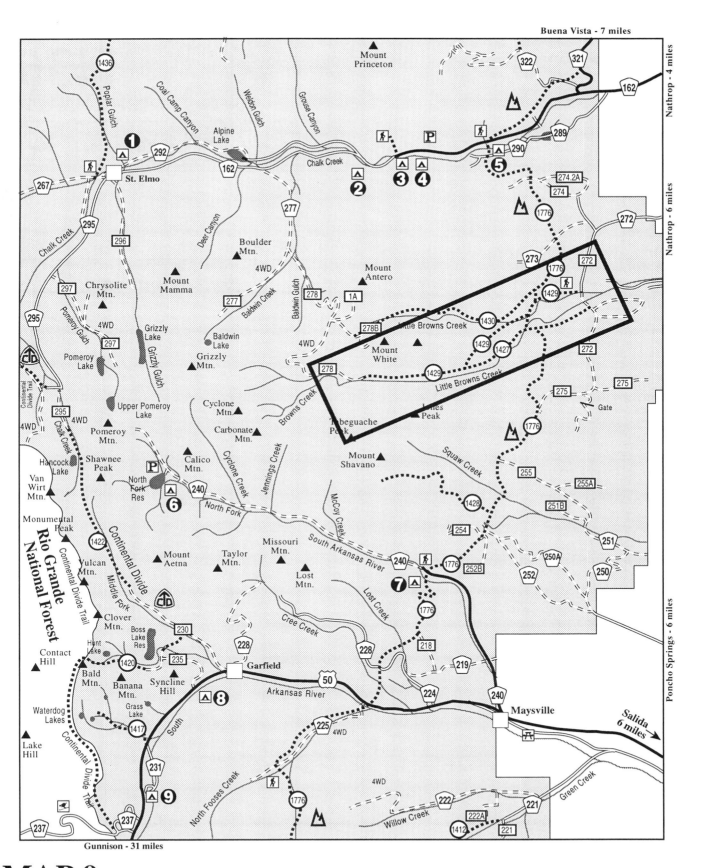

MAP 9
TRAIL #1429

© *1995 — Outdoor Books & Maps, Inc.* • *Denver, Colorado* • (303) 629-6111

ROCKY MOUNTAIN REGION

MAP NUMBER(S): 6 and 9

NATIONAL FOREST: SAN ISABEL & PIKE
RANGER DISTRICT: SALIDA

USGS: MOUNT YALE, ST. ELMO QUADS

DIFFICULTY: MORE DIFFICULT
DISTANCE: 6.0 MILES

TRAIL 1436
POPLAR GULCH

UPDATED: DECEMBER 1987

TRAIL BEGINNING ELEVATION:
10,080 ft. St. Elmo

TRAIL ENDING ELEVATION:
10,270 ft. FSR #344

USE: Moderate

RECOMMENDED SEASON: SPRING | SUMMER | FALL | WINTER

ACCESS:

#1. St Elmo: At St Elmo townsite, take County Rd. 267 (Tincup Pass Road) about 1/4 mile west to Co. Rd. 267A. Trail is about 1/10 mile in on this road.

#2. South Cottonwood Creek: From Buena Vista, take the Cottonwood Pass Road, County Road 306 about 7 miles west to County Road 344. Take Co. Rd. 344 south and west about 6 miles to the unimproved road at the trail.

No facilities are provided at the Trailheads.

ATTRACTIONS AND CONSIDERATIONS:

This is an especially scenic trail with outstanding views of the Sawatch Range and its 14000 foot peaks. Deer, elk and mountain sheep may be seen from the trail. Rocky Mountain goats also range through this area.

There are good undeveloped campsites at either end of the trail near the trail heads. Campfires are allowed as long as they are never left unattended and are thoroughly extinguished before leaving.

Motorized trailbikes are permitted on this trail. Steep grades and loose rock and gravel can be encountered on trail tread. Extra care should be taken to avoid spinning wheels and cutting ruts.

All Users, whether hiking, biking or horseback riding are requested to please respect the rights and safety of other.

NARRATIVE:

Poplar Gulch Trail runs from St. Elmo to South Cottonwood Creek.

8/18/95
tough course
1/3 - very steep dirt
1/3 - picturesque valley
1/3 - very steep again
add'l 1/2 mi @ top very steep but can see full mt. range @ top
1443 - 8/22/95
easier course
next time, go to Mount Yale
(per g.E. Jensen)

TRAIL PROFILE (ONE-WAY):

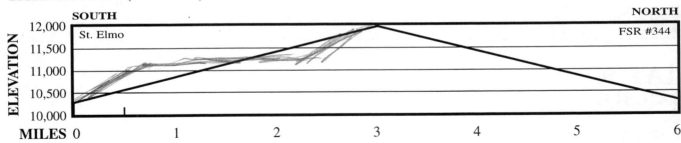

MAP 6 - South West Half

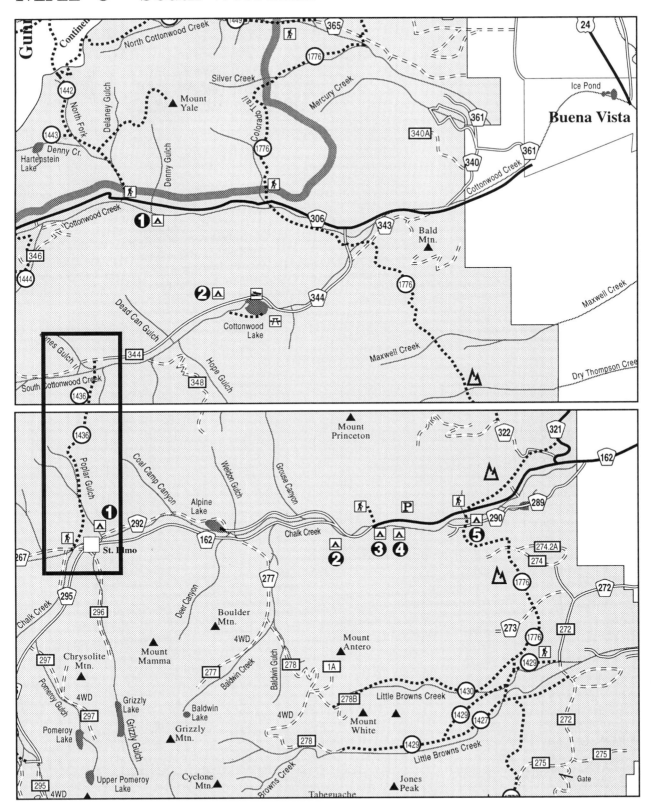

MAP 9 - North West Half
TRAIL #1436

© *1995 — Outdoor Books & Maps, Inc.* • *Denver, Colorado* • (303) 629-6111

ROCKY MOUNTAIN REGION
MAP NUMBER(S): 8

NATIONAL FOREST: SAN ISABEL
RANGER DISTRICT: SALIDA

USGS: CUMBERLAND PASS QUAD

DIFFICULTY: MORE DIFFICULT
DISTANCE: 5.0 MILES

TRAIL 1439
TUNNEL LAKE

UPDATED: SEPTEMBER 1987

TRAIL BEGINNING ELEVATION:
12,200 ft. Alpine Tunnel

TRAIL ENDING ELEVATION:
11,080 ft. FSR #267

USE: Light
No motorized vehicles

RECOMMENDED SEASON:

SPRING	SUMMER	FALL	WINTER

Across alpine meadows and willow fields, the trail is marked only with rock cairns. Use of topo map is advised.

ACCESS:
ALPINE TUNNEL: From Nathrop CO. go 15 miles west on Chalk Creek Rd. 162 to St. Elmo. Turn South on Hancock Rd No. 295 for about 5-1/2 miles then west on 4 wheel Drive Road 298, which is the old RR grade 2-1/2 miles to Alpine Tunnel. Take the trail over the tunnel to beginning of Trail 1439 at top of the pass.

TINCUP PASS ROAD: Trail begins about 4 miles west of St. Elmo on Tincup Pass ad No. 267.

There are no facilities at any of the access points.

ATTRACTIONS AND CONSIDERATIONS:
This trail offers outstanding alpine scenery in a pristine setting. Camping is permitted at undeveloped sites along the route. Best locations are at the ends of the trail. Do not use fuelwood from the timberline area.

Historic attractions in the area include St. Elmo, Romley and Hancock townsites and the Alpine Tunnel.

The trail does not go directly to Tunnel Lake.

This trail is for foot and horse only. Motor vehicles are prohibited.

Most of this trail is in alpine areas at high elevation with prolonged exposure potential severe weather. Travel is not recommended when storms threaten.

NARRATIVE:
Trail begins at saddle over Alpine Tunnel from tunnel trail No. 1438 and ends at North Fork of Chalk Creek at Tincup Pass Road.

TRAIL PROFILE (ONE-WAY):

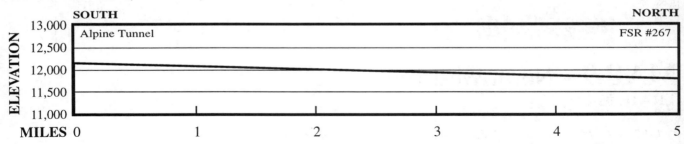

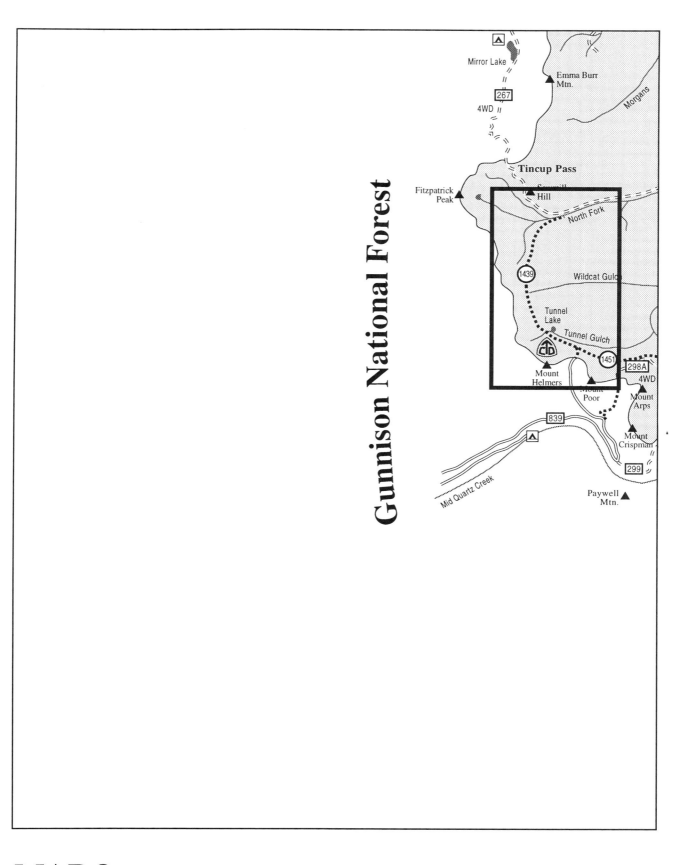

MAP 8
TRAIL #1439

ROCKY MOUNTAIN REGION

MAP NUMBER(S): 5 and 6

NATIONAL FOREST: SAN ISABEL & PIKE
RANGER DISTRICT: SALIDA

USGS: MOUNT YALE, TINCUP QUADS

DIFFICULTY: MODERATE

DISTANCE: 5.3 MILES

TRAIL 1444
PTARMIGAN LAKE

UPDATED: MARCH 1988

TRAIL BEGINNING ELEVATION:
10,670 ft. FSR #346

TRAIL ENDING ELEVATION:
12,132 ft. 4WD road on south side of Ptarmigan Lake

USE: Moderate
No motorized vehicles

RECOMMENDED SEASON:

SPRING	SUMMER	FALL	WINTER
	■	■	

ACCESS:
Ptarmigan Lake Trailhead: From Buena Vista, take Cottonwood Pass Hwy. County Road No. 306, about 14 miles west to the Ptarmigan Lake Trailhead on left, on FSR #346.

The trailhead has a vault toilet and adequate parking for 8 to 10 cars. No overnight camping facilities are provided.

Drinking water is not provided at the trailhead. Water from the creek should be treated before drinking.

ATTRACTIONS AND CONSIDERATIONS:
Ptarmigan Lake, 11 acres in size, is a popular distination for hikers. Fishing is often excellent for native cutthroat trout from the cold waters. Camping is suggested at a lower elevation below timberline. If you must gather firewood, stay as far away from the trail as possible. Do not gather brush or wood at the lake. Never cut green trees or branches for camp use. Please pack out all of your trash and a little extra if you are able. We wish you a safe and enjoyable visit.

NARRATIVE:

TRAIL PROFILE (ONE-WAY):

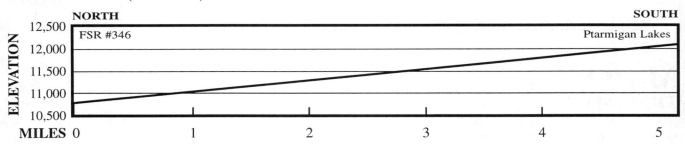

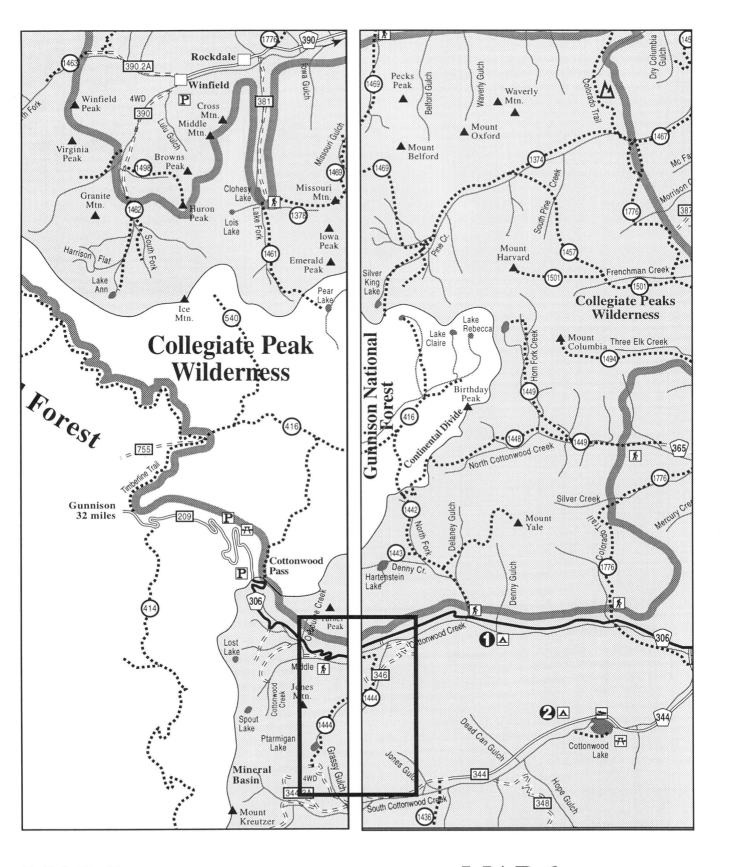

MAP 5 - East Half
TRAIL #1444

MAP 6 - West Half

© *1995 — Outdoor Books & Maps, Inc.* • *Denver, Colorado* • **(303) 629-6111**

ROCKY MOUNTAIN REGION
MAP NUMBER(S): 6

NATIONAL FOREST: SAN ISABEL
RANGER DISTRICT: SAN CARLOS

USGS: CUCHARAS PASS QUAD

DIFFICULTY: MODERATE

DISTANCE: 4.0 MILES

TRAIL 1449*
DODGETON

USE: Moderate

UPDATED:

TRAIL BEGINNING ELEVATION:
8,900 ft. Spring Creek picnic ground

TRAIL ENDING ELEVATION:
10,000 ft. Intersection of #1300

RECOMMENDED SEASON:
SPRING SUMMER FALL WINTER

ACCESS:
12 miles south of La Veta via Hwy 12 to Spring Creek Picnic Ground

ATTRACTIONS:
Scenery.

NARRATIVE:
* Trail is shown on map 22 as #1302.

TRAIL PROFILE (ONE-WAY):

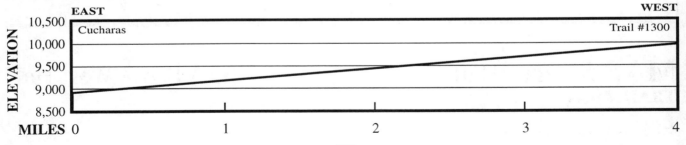

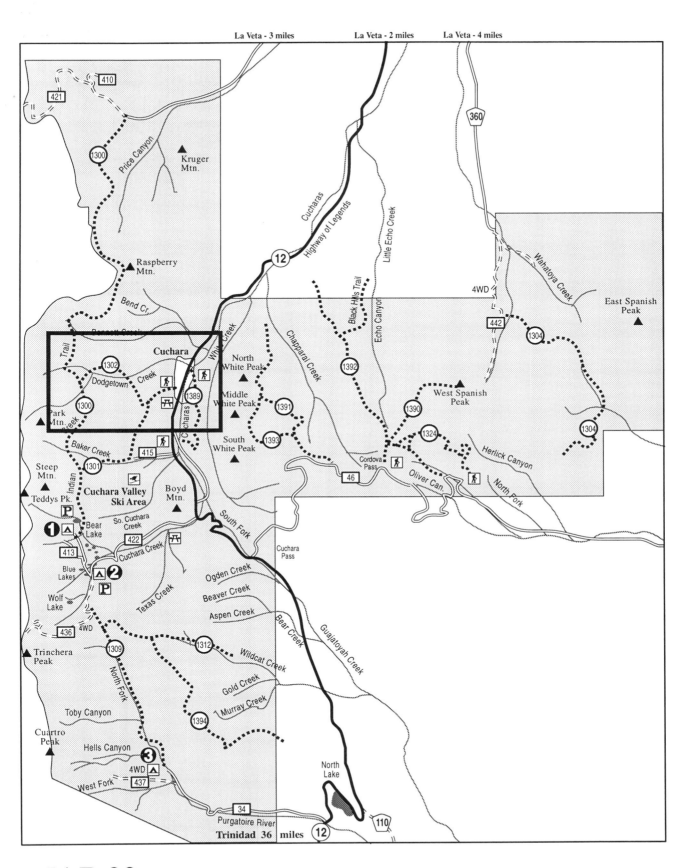

MAP 22
TRAIL #1449

© 1995 — Outdoor Books & Maps, Inc. • Denver, Colorado • (303) 629-6111

ROCKY MOUNTAIN REGION
MAP NUMBER(S): 5 AND 6

NATIONAL FOREST: SAN ISABEL & PIKE
RANGER DISTRICT: LEADVILLE

USGS: WINFIELD QUADS

DIFFICULTY: MORE DIFFICULT
DISTANCE: 6.5 MILES

TRAIL 1469
MISSOURI GULCH

UPDATED: MARCH 1988

TRAIL BEGINNING ELEVATION:
9,950 ft. Trailhead on County Road #390

TRAIL ENDING ELEVATION:
11,600 ft. Intersection Trail #1374

ACCESS:
Trailhead at Vicksburg approximately 8 miles west of U.S. highway 24 on Forest Road #120 on Clear Creek. Clear Creek is about 19 miles south of Leadville or 14 miles north of Buena Vista, Colorado. Forest Road #120 is graveled and sometimes rough but satisfactory for most vehicles.

The trailhead provides parking space but has no restrooms, water or overnight facilities.

ATTRACTIONS:
This trail is for foot and horse travel only. Motorized travel is prohibited.

Trail begins at Vicksburg Townsite on Clear Creek and ends at Pine Creek Trail in Missouri Basin on Pine Creek.

USE: Heavy
No motorized vehicles

RECOMMENDED SEASON:

| SPRING | SUMMER | FALL | WINTER |

NARRATIVE:

TRAIL PROFILE (ONE-WAY):

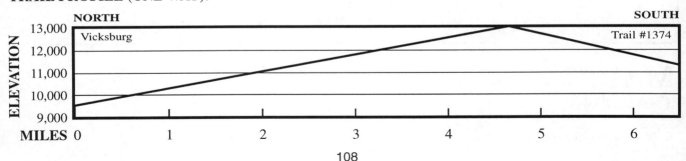

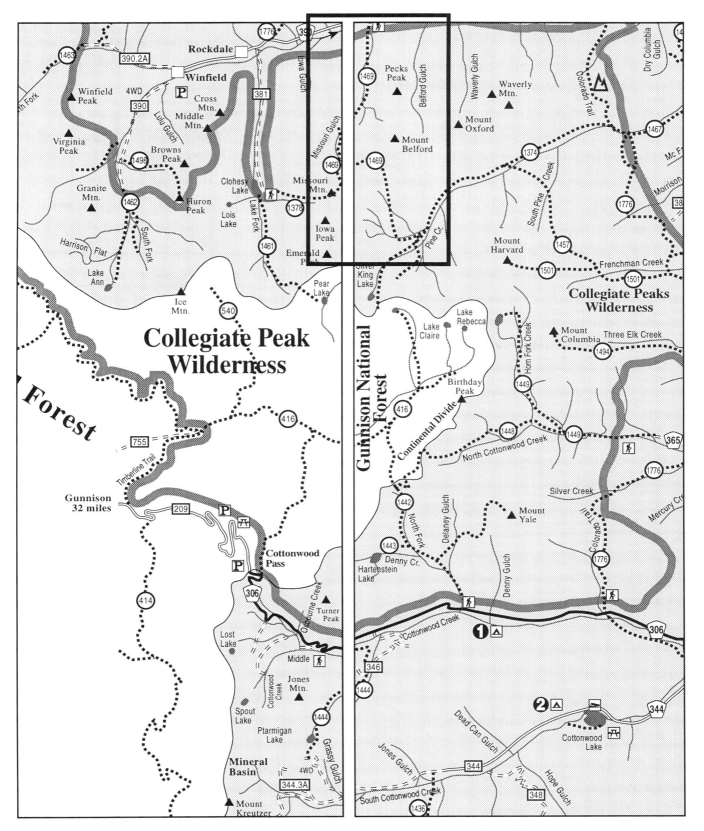

MAP 5 - West Half   **MAP 6** - East Half

TRAIL #1469

© 1995 — *Outdoor Books & Maps, Inc.* • *Denver, Colorado* • (303) 629-6111

ROCKY MOUNTAIN REGION
MAP NUMBER(S): 3

NATIONAL FOREST: SAN ISABEL & PIKE
RANGER DISTRICT: LEADVILLE

USGS: MT. ELBERT, MT. HARVARD QUADS

DIFFICULTY: MODERATE TO DIFFICULT
DISTANCE: 3.5 MILES

TRAIL 1474
LA PLATA PEAK

UPDATED: DECEMBER 1990

TRAIL BEGINNING ELEVATION:
10,150 ft. Colorado 82 on South Fork Lake Creek

TRAIL ENDING ELEVATION:
14,336 ft. La Plata Peak

USE: Moderate

RECOMMENDED SEASON:

SPRING	SUMMER	FALL	WINTER

ATTRACTIONS AND CONSIDERATIONS:
No well defined or marked trail exists above treeline.

NARRATIVE:

ACCESS:
ROUTE 1
From Leadville travel south on US Hwy 24 to Balltown, then west on Colo. 82 past Twin Lakes to S. Fork Lake Creek.

Trail begins at Winfield, 4-WD may drive 1-1/2 miles (N. Fork Clear Creek) to road which forks north, otherwise may walk to this point; head north through aspen clone. Just before gate, notice cairn marking path which continues north. Path soon crosses creek. Continue on path approximately 1 3/4 miles to saddle above treeline. View to north is over-looking La Plata Gulch. Continue northeast up ridge to summit, approximately 1-1/2 miles. Trail ends at summit of La Plata Peak, 5th highest in the state.

ROUTE 2
From Leadville travel south on US Hwy 24 to Clear Creek, County Road 390. Travel W on Road 120 to Winfield, approximately 12 miles from the highway.

Trail begins at South Fork Lake Creek bridge, cross Lake Creek on bridge. Walk path paralleling creek east approximately 1 mile to aspen clone and meadows where cairns show location of trail forking SW. Cross La Plata Gulch drainage, continue S approximately 1/2 mile to glacial basin. Continue to approximately elevation of 11,400 feet, fork E to attain ridge at saddle approximately 1 mile, 1,250' vertical, continue S on ridge to summit, approximately 1 mile, 1,700' vertical. Trail ends at summit of La Plata Peak.

TRAIL PROFILE (ONE-WAY):

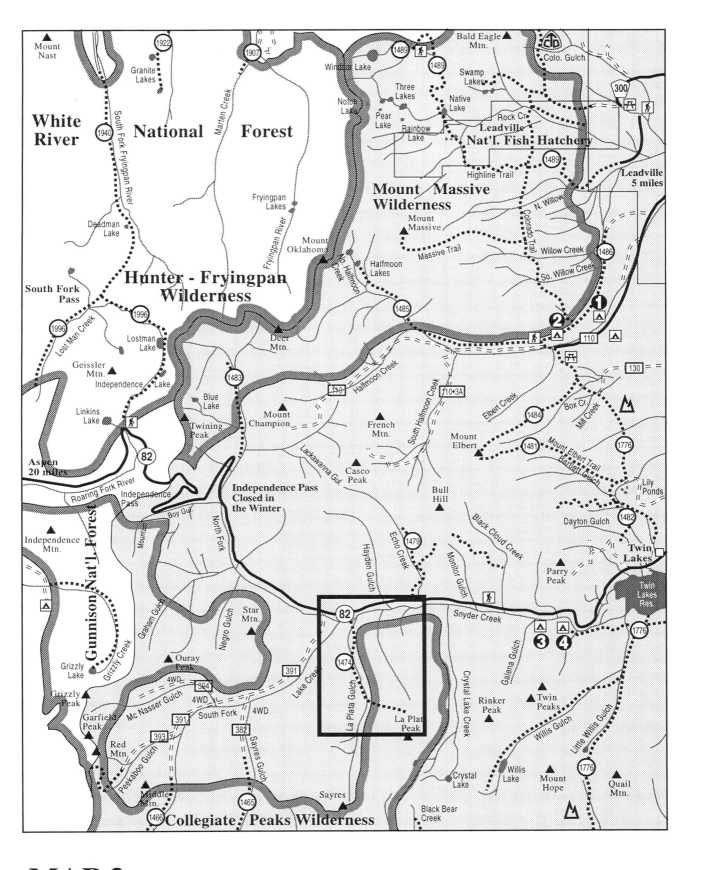

MAP 3
TRAIL #1474

© 1995 — *Outdoor Books & Maps, Inc.* • *Denver, Colorado* • (303) 629-6111

ROCKY MOUNTAIN REGION
MAP NUMBER(S): 3

NATIONAL FOREST: SAN ISABEL & PIKE
RANGER DISTRICT: LEADVILLE

USGS: MT. ELBERT QUAD

DIFFICULTY: MORE DIFFICULT
DISTANCE: 6.0 MILES

TRAIL 1481
SOUTH MT. ELBERT

UPDATED: MAY 1988

TRAIL BEGINNING ELEVATION:
9,622 ft. Lily ponds

TRAIL ENDING ELEVATION:
14,433 ft. South side of Mt. Elbert

USE: Heavy
No motorized vehicles

RECOMMENDED SEASON:

SPRING	SUMMER	FALL	WINTER

ACCESS:
From Leadville travel south on US Hwy 24 to the junction of Colorado Hwy 82. Continue approximately 3-3/4 miles west to the Lakeview Campground turnoff, north on Lake County Rd. 24. Continue beyond the campground approximately 1/4 mile to trailhead where parking is available. The trail will follow the Colorado Trail No. 1776 for approximately three miles and then veer NW onto the South Mt. Elbert Trail No. 1481 to the summit.

ATTRACTIONS AND CONSIDERATIONS:
The Colorado Trail No. 1776 and both Mt Elbert Trails No.'s 1481 and 1484 are for foot and horse travel only. Motorized use is prohibited.

NARRATIVE:

TRAIL PROFILE (ONE-WAY):

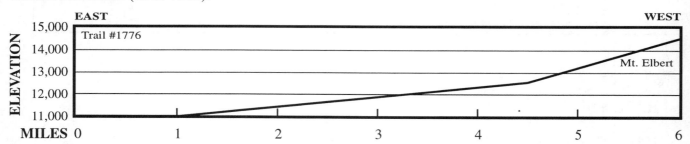

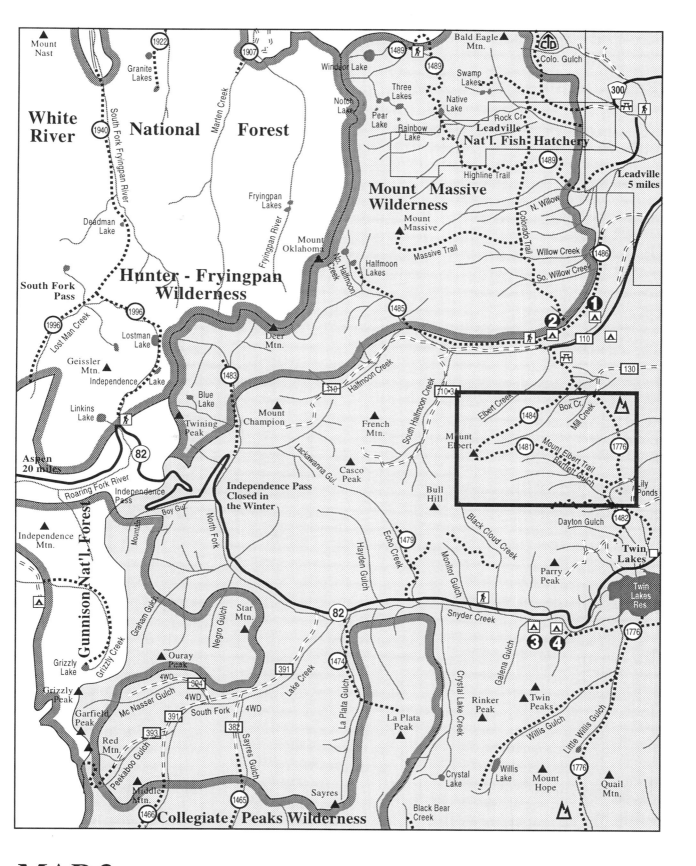

MAP 3
TRAIL #1481

© 1995 — Outdoor Books & Maps, Inc. • Denver, Colorado • (303) 629-6111

ROCKY MOUNTAIN REGION
MAP NUMBER(S): 3

NATIONAL FOREST: SAN ISABEL & PIKE
RANGER DISTRICT: LEADVILLE

USGS: MT. ELBERT, MT. MASSIVE QUADS

DIFFICULTY: MOST DIFFICULT
DISTANCE: 4.5 MILES

TRAIL
1484
NORTH MT. ELBERT

UPDATED: MARCH 1988

TRAIL BEGINNING ELEVATION:
10,100 ft. FSR #110

TRAIL ENDING ELEVATION:
14,433 ft. Mt. Elbert

USE: Very heavy
No motorized vehicles

RECOMMENDED SEASON:

| SPRING | SUMMER | FALL | WINTER |

ACCESS:
Access to the trailhead on Halfmoon Creek beginning from Leadville is as follows: Travel three (3) miles southwest on US Hwy 24 to County Highway 300. Then approximately 0.8 mile west to Forest Road 110. On Forest Road 110 travel south about 7 miles to the Trailhead

ATTRACTIONS AND CONSIDERATIONS:
The Colorado Trail No. 1776 and both Mt. Elbert Trails 1376 and 1484 are for foot and horse travel only.

Motorized use is prohibited.

NARRATIVE:

TRAIL PROFILE (ONE-WAY):

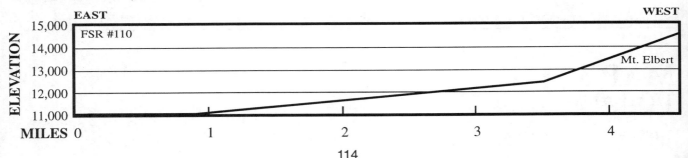

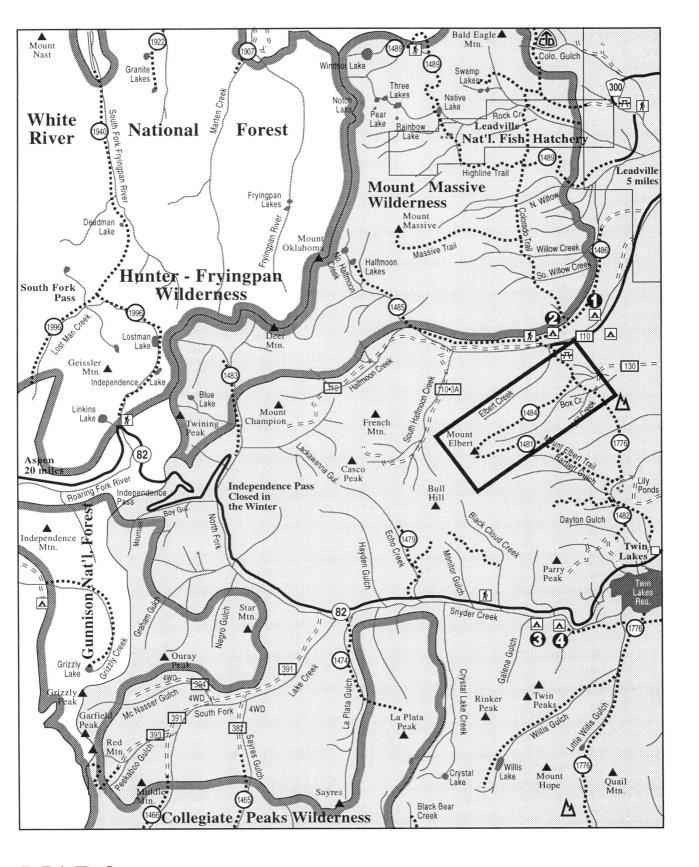

MAP 3
TRAIL #1484

© 1995 — *Outdoor Books & Maps, Inc.* • *Denver, Colorado* • *(303) 629-6111*

ROCKY MOUNTAIN REGION
MAP NUMBER(S): 6

NATIONAL FOREST: SAN ISABEL & PIKE
RANGER DISTRICT: LEADVILLE

USGS: MT. HARVARD QUAD

DIFFICULTY: MODERATE TO DIFFICULT
DISTANCE: 7.5 MILES ROUTE 2

TRAIL 1494
MOUNT COLUMBIA

UPDATED: DECEMBER 1990

TRAIL BEGINNING ELEVATION:
8,400 ft. FSR #368

TRAIL ENDING ELEVATION:
14,073 ft. Summit of Mount Columbia

USE: Moderate

RECOMMENDED SEASON: SPRING | SUMMER | FALL | WINTER

ACCESS:
ROUTE 1*
From Leadville travel south on US Hwy 24 for 26.5 miles, drive west into the Four Elk development and follow the signs to the forest access road. Please respect private property.

Where trail begins parking is available at beginning of forest access road between Three Elk and Four Elk. 4WD may continue one mile to road closure. Continue up old road one mile to Colorado Trail. Hike one mile south on Colorado Trail to Three Elk Trail. Hike west on Three Elk to Columbia Basin 2.5 miles. Climb NW 0.5 mile up gully to ridge and continue 0.5 miles SW to summit. Trail ends at summit of Mount Columbia

ROUTE 2*
From Leadville, travel south 24 miles on US Hwy 24 to wilderness boundary, travel SW 3.3 miles to Forest Road 386.

Where trail begins parking is available at beginning of Harvard Trail, Forest Road 386. 4WD may continue 3.5 miles to Wilderness boundary. From there, continue one mile to Colorado Trail and continue up the Frenchman Creek Drainage 2 miles. Head SW up gentle slope one mile to summit. Trail ends at summit of Mount Columbia

* NOTE: No well-defined or marked trail exists above treeline on either route to summit.

ATTRACTIONS:

NARRATIVE:

TRAIL PROFILE (ONE-WAY):

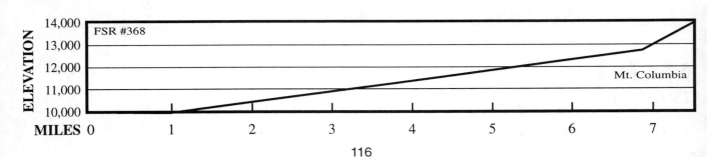

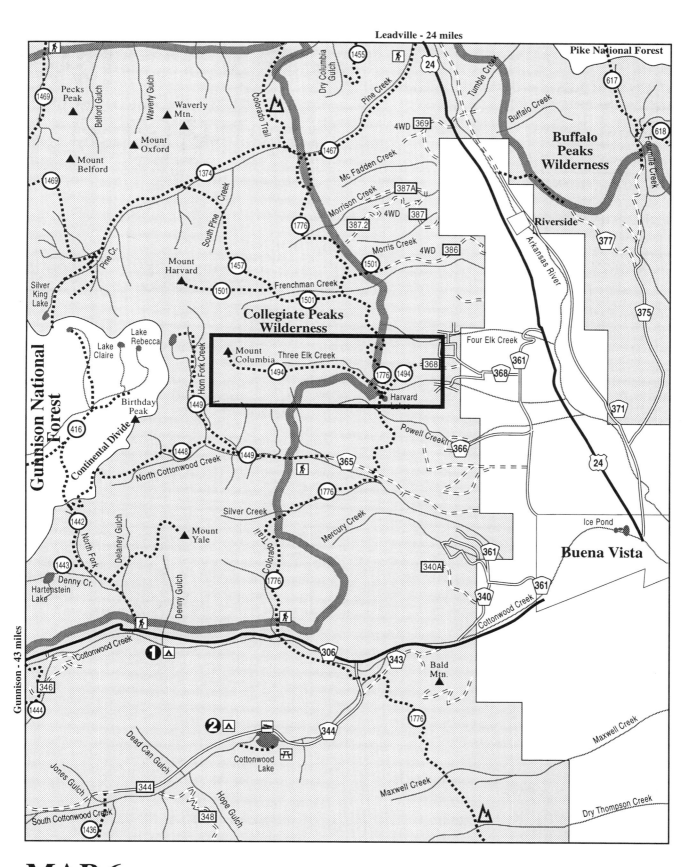

MAP 6
TRAIL #1494

© 1995 — Outdoor Books & Maps, Inc. • Denver, Colorado • (303) 629-6111

ROCKY MOUNTAIN REGION
MAP NUMBER(S): 5

NATIONAL FOREST: SAN ISABEL & PIKE
RANGER DISTRICT: LEADVILLE

USGS: MT. HARVARD QUAD

DIFFICULTY: VERY DIFFICULT
DISTANCE: 3.5 MILES

TRAIL 1498
MOUNT HURON (ROUTE 2)

UPDATED: DECEMBER 1990

TRAIL BEGINNING ELEVATION:
10,400 ft. FSR #390

TRAIL ENDING ELEVATION:
14,005 ft. Summit of Mount Huron

USE: Moderate to Heavy
No motorized vehicles

RECOMMENDED SEASON:
SPRING | **SUMMER** | FALL | WINTER

ACCESS:
From Leadville travel south on US Hwy 24 to Granite, continue 1-1/2 miles to Clear Creek Road, Forest Road 390, Chaffee County. Travel west 12 miles to Winfield then approximately 1/2 mile south on Forest Road 390 (S. Fork) to park.

ATTRACTIONS AND CONSIDERATIONS:
Travel past road closure limited to foot & horse travel.

No motorized travel.

NARRATIVE:
ROUTE 1
Parking is available approximately 1/2 mile south of Winfield on South Fork of Clear Creek Road, Forest Road 390. 4WD may continue 1-1/2 miles to Wilderness boundary. Otherwise walk road to Wilderness boundary, continue 2 miles more to meadows across creek from site of Hamilton. Take South (left) trail fork switchback to site of Wallace Mine at treeline. Continue NE up SW ridge to summit, exercising caution on loose rocks. Trail ends at summit of Mount Huron, 52nd highest in the State.

ROUTE 2
Parking is available approximately 1/2 mile South of Winfield on South Fork of Clear Creek Rd, Forest Road 390. 4WD may continue 1-1/2 miles to Wilderness boundary. From there, climb ESE up steep open meadow to cirque between Browns Peak and Huron Peak. From cirque head SW up boulder field to summit, exercising caution on loose rocks. Trail ends at summit of Mount Huron.

TRAIL PROFILE (ONE-WAY):

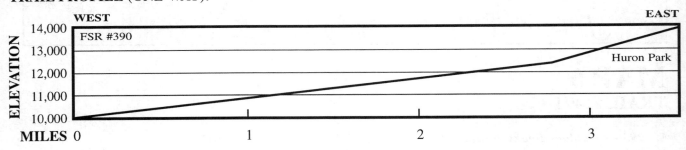

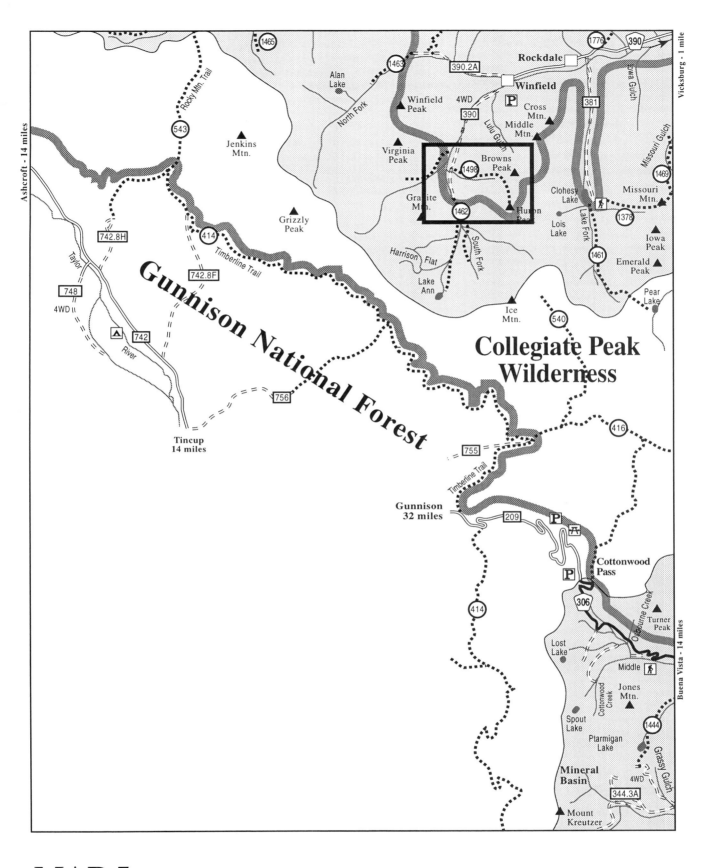

MAP 5
TRAIL #1498

© 1995 — Outdoor Books & Maps, Inc. • Denver, Colorado • (303) 629-6111

ROCKY MOUNTAIN REGION
MAP NUMBER(S): 6

NATIONAL FOREST: SAN ISABEL & PIKE
RANGER DISTRICT: LEADVILLE

USGS: MT. HARVARD QUAD

DIFFICULTY: LOW TO MODERATE
DISTANCE: 8.5 MILES

TRAIL 1501
MT. HARVARD

UPDATED: DECEMBER 1990

TRAIL BEGINNING ELEVATION:
8,400 ft. FSR #386 for Route 1 and 9,100 ft. for Route 2

TRAIL ENDING ELEVATION:
14,420 ft. Summit Mount Harvard

ACCESS:
#1: From Leadville travel 24 miles to Chaffee County Route 386. Travel SW 1.5 miles to Forest Road 368.

#2: From Leadville travel south on US Hwy 24 to Granite. Continue 4 miles south to Cold Camp. Turn off highway 0.25 miles past Cold Camp onto Forest Road 388.

ATTRACTIONS AND CONSIDERATIONS:
The Trail is a popular hiking trail which terminates at the lake. The lake is nestled in a high mountain cirque between two mountains. The rocky slopes surrounding the lake are home to bighorn sheep and mountain goats. The trail is easy to follow and is signed at appropriate trail junctions. The elevation gain to the lake is 3,030 feet. The final two miles up to the lake are above treeline and very picturesque, especially in the fall.

USE: Route 1 - Moderate
Route 2 - High

RECOMMENDED SEASON:

SPRING	SUMMER	FALL	WINTER

NARRATIVE:
ROUTE 1*
(Elev. 8,400 ft.) Parking is available at beginning of Harvard Trail, Forest Road 386. 4WD may continue 3.5 miles to wilderness boundary. From there continue 1 mile to Colorado Trail, Forest Trail 1776. Cross the Colorado Trail and continue up the Frenchman Creek drainage 2.5 miles. Head WNW 1.5 miles up the ridge to the summit. Trail ends at summit of Mount Harvard, 3rd highest in the state.

ROUTE 2*
(Elev. 9,100 ft.) Parking is available on Pine Creek Road, Forest Road 388. 4WD may continue 1 mile to road closure. From there follow Pine Creek Trail 2.5 miles to th Colorado Trail. Cross the bridge and continue up the valley 2.5 miles to Little John's Cabin. From Little John's Cabin cross the creek to the forge and head SE 1 mile on the South Pine Creek Trail to the top of the ridge. 2 miles SW to the summit. Trail ends at summit of Mount Harvard, 3rd highest in the state.

*NOTE: No well defined or marked trail exists above treeline on either route to summit.

TRAIL PROFILE (ONE-WAY):

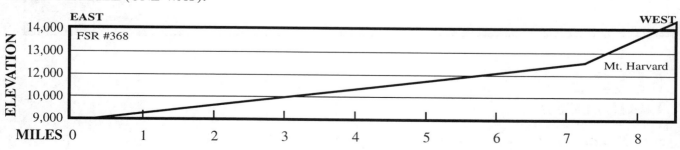

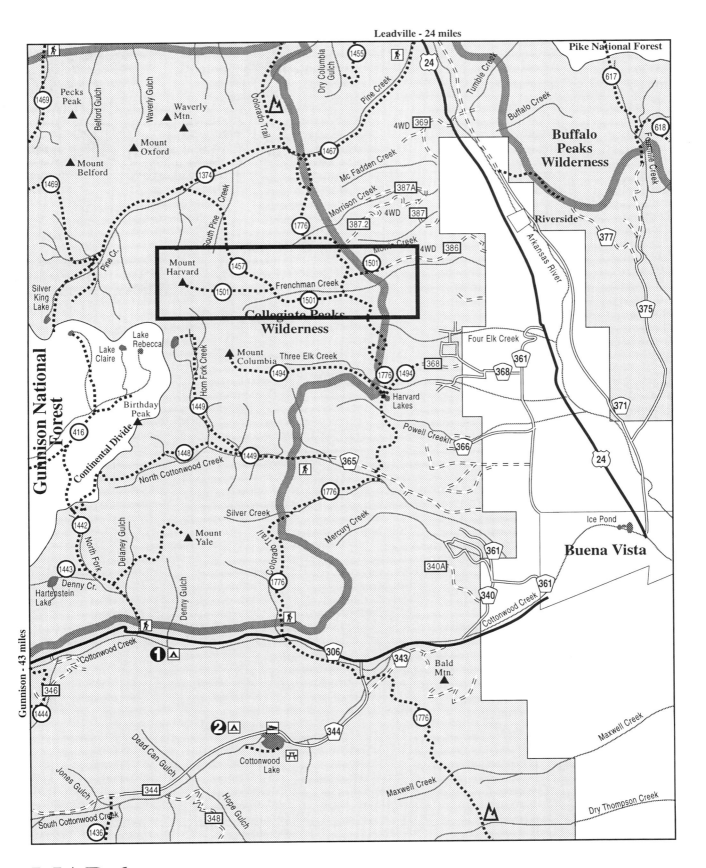

MAP 6
TRAIL #1501

© 1995 — *Outdoor Books & Maps, Inc.* • *Denver, Colorado* • (303) 629-6111

ROCKY MOUNTAIN REGION

MAP NUMBER(S): 1 and 2

NATIONAL FOREST: SAN ISABEL
RANGER DISTRICT: LEADVILLE

USGS: LEADVILLE NORTH, HOMESTAKE RESERVOIR QUADS

UPDATED: MARCH 1988

TRAIL BEGINNING ELEVATION:
10,480 ft.

TRAIL ENDING ELEVATION:
9,869 ft.

ACCESS:

#1. Tennessee Pass: Off of Hwy. 24. Across the road from the 10th Mountain Division monument and Ski Cooper.

#2. Turquoise Lake at Lake Fork: From center of Leadville, turn onto West 6th at traffic light for 1/2 mile. Turn right onto County Rd. 4 for 5 miles to the Dam and continue on County Rd. 4 for 5 more miles to Lake Fork near the May Queen Campground.

ATTRACTIONS AND CONSIDERATIONS:

Tennessee Pass offers trailhead parking, restrooms, but no water. This segment of the trail offers various life zones from marshy areas to high alpines including lodgepole pines to spruce/fir trees. There are many high alpine lakes found in this segment of the Trail. Progressing southward along Longs Gulch, it crosses the Wurtz Ditch, which leads to one of many high lakes found in the Holy Cross Wilderness, Slide Lake. Also found southward are Deckers Lake, St. Kevin's Lake, Bear Lake, and Galena Lake. These trails make nice side trips off the trail, but require topographical maps. Two of the largest peaks are found in this section. They are Homestake (13,209 ft.) and Galena (12,893 ft.) Peaks. Snow is expected to be encountered until mid-July. You will be traveling in and out of the Holy Cross Wilderness. The following regulations should be observed:
1. Dogs must be on a leash.
2. Party limit of 25 people.
3. Camping within 100 feet of Bear Lake, St. Kevins's Lake, Timberline Lake and West Tennessee Lake is prohibited.

DIFFICULTY: MODERATE

DISTANCE: 12.5 MILES

TRAIL 1776A-B COLORADO*

USE: Light

RECOMMENDED SEASON:

SPRING	SUMMER	FALL	WINTER
	■	■	

* COLORADO TRAIL
TENNESSEE PASS TO TURQUOISE LAKE

TRAIL BEGINS: (This segment) Tennessee Pass on State Hwy. 24
Elevation: 10,480 ft. (3.2 km)

TRAIL ENDS: (This segment) Turquoise Lake at Lake Fork (near May Queen Campground)
Elevation: 9,869 ft. (2.8 km)

TRAIL PROFILE (ONE-WAY):

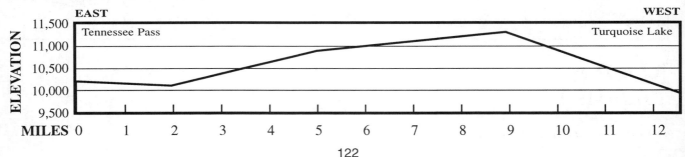

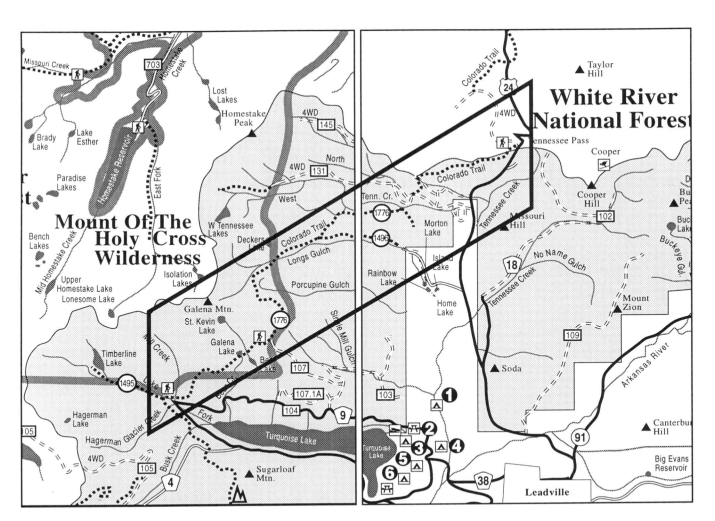

MAP 1 - South East Half
TRAIL #1776A-B

MAP 2 - South West Half

© 1995 — *Outdoor Books & Maps, Inc.* • Denver, Colorado • **(303) 629-6111**

ROCKY MOUNTAIN REGION

MAP NUMBER(S): 1 and 3

NATIONAL FOREST: SAN ISABEL
RANGER DISTRICT: LEADVILLE

USGS: MT. MASSIVE, HOMESTAKE RESERVOIR QUADS

UPDATED: MARCH 1988

TRAIL BEGINNING ELEVATION:
9,869 ft. Turquoise Lake

TRAIL ENDING ELEVATION:
10,100 ft. Halfmoon Creek

ACCESS:

#1. Turquoise Lake: From center of Leadville, turn onto West 6th for 1/2 mile. Turn right onto County Rd. 4 for 5 miles to the Dam and continue on County Rd. 4 for 5 more miles to Lake Fork near the May Queen Campground.

#2. Halfmoon Creek: Travel 3 miles SW of U.S. Hwy. 24 to Hwy. 300, then approximately .8 mile to Forest Rd. 110. Travel S about 7 miles to the trailhead past Elbert Creek Cmpgrd.

ATTRACTIONS AND CONSIDERATIONS:

On this segment of the Trail, there are numerous streams coming off of the Continental Divide, many offering good fishing are encountered. Proceeding south is Busk Creek and also the Hagerman Pass Road, which was the route to the highest trans-continental standard gauge railroad. This is now a 4-wheel drive road. Sugarloaf Mountain, which was a primary mineral district of turn-of-the-century mining is an added attraction. At this point, Turquoise Lake, a man-made reservoir and Catchmen's basin of the Frying Pan-Arkansas project can be observed. Mt. Massive Trail, which leads to Colorado's second highest peak, Mt. Massive (14,421 ft.) is intersected along the trail. Dropping down is the Halfmoon Campground, ($6.00/nIght) which provides individual camping units, trash disposal, restrooms, and water. You will be traveling in and out of the Mt. Massive Wilderness. The following rules should be observed:
1. Dogs must be on a leash.
2. Party limit of 25 people.
3. Do not camp within 100 ft. of Native or Windsor Lakes.

TRAIL

DIFFICULTY: MODERATE
DISTANCE: 13.0 MILES

1776B-C
COLORADO*

USE: Heavy

RECOMMENDED SEASON:

| SPRING | SUMMER | FALL | WINTER |

* COLORADO TRAIL
TURQUOISE LAKE TO HALFMOON CREEK

TRAIL BEGINS: (This Segment) Turquoise Lake at Lake Fork (near May Queen Campground)
Elevation: 9,869 ft. (2.8 km.)

TRAIL ENDS: (This Segment) Halfmoon Creek at County Rd. 11 (near Halfmoon Campground)
Elevation: 10,110 ft. (3.2 km.)

TRAIL PROFILE (ONE-WAY):

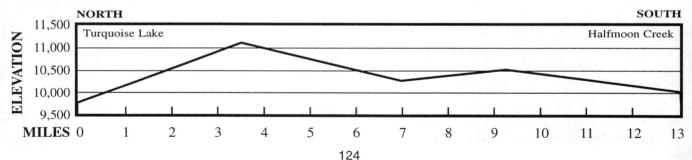

MAP 1 - South Half

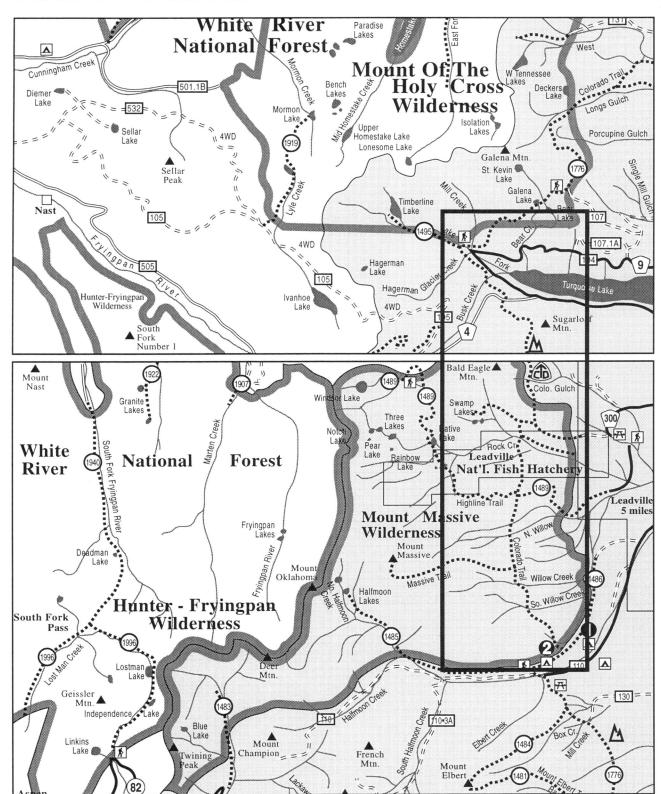

MAP 3 - North Half
TRAIL #1776B-C

© *1995 — Outdoor Books & Maps, Inc.* • *Denver, Colorado* • (303) 629-6111

ROCKY MOUNTAIN REGION

MAP NUMBER(S): 3

NATIONAL FOREST: SAN ISABEL
RANGER DISTRICT: LEADVILLE

USGS: MOUNT MASSIVE, MOUNT ELBERT QUADS

DIFFICULTY: MODERATE

DISTANCE: 8.0 MILES

TRAIL 1776C-D
COLORADO*

UPDATED: MARCH 1988

TRAIL BEGINNING ELEVATION:
10,100 ft. Halfmoon Creek

TRAIL ENDING ELEVATION:
9,200 ft. Twin Lakes

USE: Heavy

RECOMMENDED SEASON: SPRING **SUMMER FALL** WINTER

ACCESS:

#1. Halfmoon Creek: Travel 3 miles SW of U.S. Hwy. 24 to Hwy. 300. Then approximately 0.8 mile to Forest Rd. 110. Travel south about 7 miles to the Trailhead past Elbert Creek Campground.

#2. Twin Lakes: From Leadville, travel south on U.S. Hwy. 24 to Colo. Hwy. 82.

#3. Lakeview Campground. At the Twin Lakes Recreation Area on U.S. 24, then right on State Hwy. 82.

ATTRACTIONS AND CONSIDERATIONS:

On this segment of the trail, there is an intersection of the trail which leads to the highest peak in the state, Mt. Elbert (14,433 ft.). There will be variations in vegetation and ecological systems all along the Trail. The southern portion of this segment of the trail is a winter elk range, where approximately 200 elk may be seen upon reaching Twin Lakes. The Mt. Elbert Trail is once again intersected. The trail will follow a 4WD road past some old mining remains and on into Lakeview Campground. The trail will continue under Hwy. 82, past the power plant to the Twin Lakes Dam. One remaining cabin on the lake, Cabin Cove, which is a group reservation camping site, can be observed.

* COLORADO TRAIL
HALFMOON CREEK TO TWIN LAKES

TRAIL BEGINS: (This Segment) Halfmoon Creek at County Rd. 11 (near Halfmoon Campground).
Elevation: 10,100 ft. (3.2 km.)

TRAIL ENDS: (This Segment) Loop H (Campground Facilities) or Twin Lakes at the Dam.
Elevation: 9,200 ft. (2.7 km.)

TRAIL PROFILE (ONE-WAY):

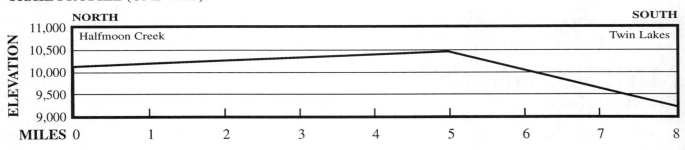

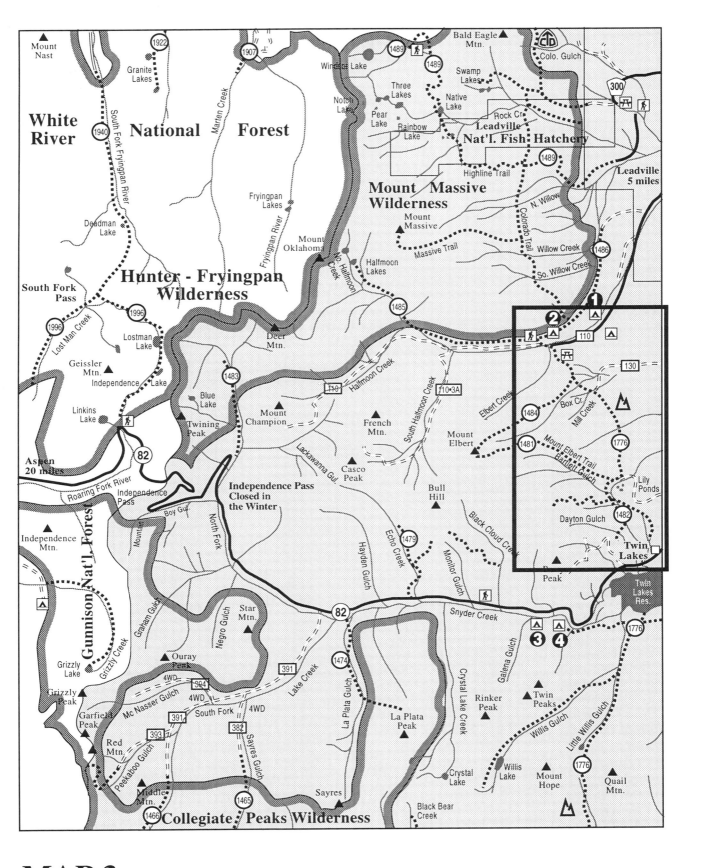

MAP 3
TRAIL #1776C-D

ROCKY MOUNTAIN REGION

MAP NUMBER(S): 4 and 5

NATIONAL FOREST: SAN ISABEL
RANGER DISTRICT: LEADVILLE

USGS: MOUNT ELBERT, GRANITE QUADS

DIFFICULTY: DIFFICULT

DISTANCE: 20.0 MILES

TRAIL
1776D-E
COLORADO*

UPDATED: MARCH 1988

TRAIL BEGINNING ELEVATION:
9,200 ft. Twin Lakes

TRAIL ENDING ELEVATION:
8,940 ft. Cty. Rd. #390

USE: Light

RECOMMENDED SEASON:

SPRING	SUMMER	FALL	WINTER

ACCESS:

#1. Twin Lakes: From Leadville, travel south on U.S. Hwy 24 to Colo. Hwy 82. Continue 3/4 mile to the Lakeview Campground, Loop H or 3 miles east to the Twin Lakes Dam.

#2. Clear Creek: At 2 miles south of Granite on U. S. Hwy. 24, turn west on County Road 390 for 3 miles where the trail Crosses the road.

ATTRACTIONS AND CONSIDERATIONS:

This segment of the Trail begins at Loop H at the Lakeview Campground, which drops down to an underpass under Hwy. 82. There are several other units at this campground ($6.00/night). One interesting site that may be observed is the Bureau of Reclamation's hydro-electric plant. The information center, because of its professional displays, should be visited. Continuing west at close proximity are the Lakes; which offer good fishing. Along the shore are the Dexter Boat Ramp, picnic ground and Campground ($6.00/night). On the south side or the lakes at the isthmus of the two lakes is the historic site of Interlaken, where interpretive signs denote an 18th Century Hotel. For those who have previously hiked this Trail, right-of-way complications have caused the trail to now go over Hope Pass. While climbing Hope Pass, (12,500 ft.), adverse weather conditions and steep climbing toward Clear Creek Canyon are encountered. Southward is the historic town of Vicksburg, whose museum and livery stable are an added attraction.

* COLORADO TRAIL
TWIN LAKES TO CLEAR CREEK

TRAIL BEGINS: (This Segment) Twin Lakes at the Dam
Elevation: 9,200 ft. (2.7 km)

TRAIL ENDS: (This Segment) Clear Creek at Forest Rd. 306
Elevation: 8,940 ft. (2.6 km)

TRAIL PROFILE (ONE-WAY):

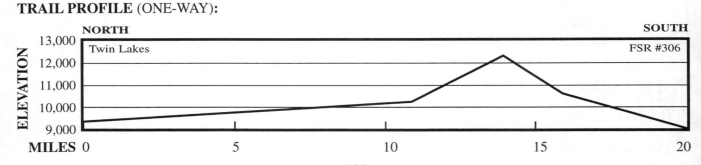

MAP 3 - South East

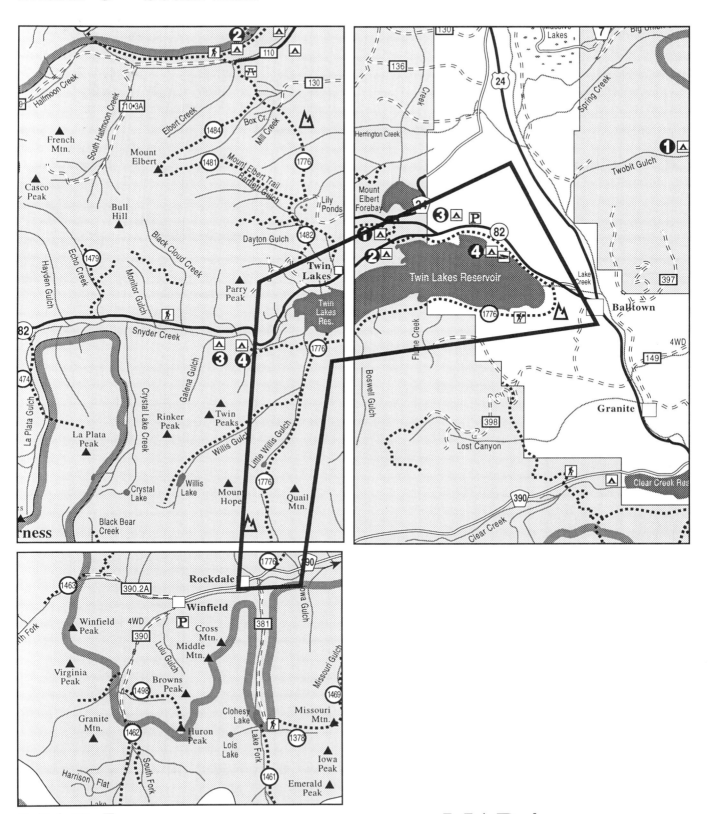

MAP 5 - North East
TRAIL #1776D-E

MAP 4 - South West

© 1995 — *Outdoor Books & Maps, Inc.* • *Denver, Colorado* • (303) 629-6111

ROCKY MOUNTAIN REGION

MAP NUMBER(S): 4 and 6

NATIONAL FOREST: SAN ISABEL
RANGER DISTRICT: SALIDA

USGS: GRANITE, MOUNT HARVARD, HARVARD LAKE, BUENA VISTA WEST

UPDATED: MARCH 1988

TRAIL BEGINNING ELEVATION:
8,960 ft. Cty. Rd #390

TRAIL ENDING ELEVATION:
9,350 ft. N. Cottonwood Creek

ACCESS:

#1. Clear Creek: At 2 miles south of Granite on US Hwy 24, turn west on County Road 390 for 3 miles where the trail crosses the road.

#2. North Cottonwood Creek: From US Hwy 2 at the north end of Buena Vista, take County Road 350 to the west about 2 miles to County Road 261. Turn north for about 1 mile to County Road 365 on left. Follow Road 365 about 3-1/2 miles to trail. Road 365 is very rough' and is recommended for high clearance vehicles only.

Trailheads have limited space for parking. Toilets, drinking water or other facilities are not provided.

ATTRACTIONS AND CONSIDERATIONS:

Collegiate Peaks Wilderness: Much of this segment of the trail passes through the Collegiate Peaks Wilderness. To protect the qualities for which the Wilderness was established and to increase your enjoyment, certain regulations and restrictions are enforced in the area. Mechanized transportation and equipment including motorcycles, mountain bicycles and hang gliders are not allowed. Possession of power saws and other motorized equipment is prohibited as well. Dogs must be kept on a six foot leash. To provide solitude. the maximum size of groups is restricted to 25 persons and/or pack or saddle stock. Special permits are not required.

Camping is allowed at undeveloped sites along the trail. The Pine Creek area and Rainbow and Harvard Lakes vicinities receive very heavy use. Camping is not recommended in those areas during heavy use seasons. Camping at Clear Creek is recommended at the State developments at Clear Creek Reservoir 1 mile to the east or south of Clear Creek after passing through the private lands. Please obey all signs to avoid trespassing.

TRAIL 1776E-F COLORADO*

DIFFICULTY: MODERATE TO MORE DIFFICULT
DISTANCE: 15.5 MILES

USE: Heavy
No motorized vehicles, bikes or hang gliders

RECOMMENDED SEASON:
SPRING SUMMER FALL WINTER

* COLORADO TRAIL
CLEAR CREEK TO N. COTTONWOOD CREEK

TRAIL BEGINS: (This Segment) Clear Creek at County Rd. 390.
Elevation: 8,960 ft. (2727km)

TRAIL ENDS: (This Segment) North Cottonwood Creek at Co. Rd. 365.
Elevation: 9,350 ft. (2856km)

TRAIL PROFILE (ONE-WAY):

MAP 4 - South West Quarter

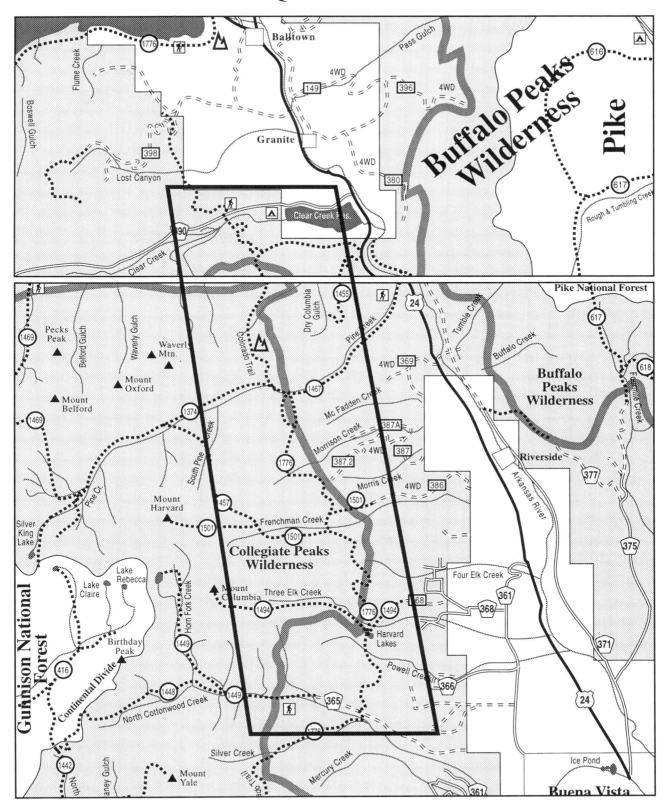

MAP 6 - North West Three Quarters
TRAIL #1776E-F

ROCKY MOUNTAIN REGION
MAP NUMBER(S): 6

NATIONAL FOREST: SAN ISABEL
RANGER DISTRICT: SALIDA

USGS: MOUNT YALE, BUENA VISTA WEST QUADS

UPDATED: MARCH 1988

TRAIL BEGINNING ELEVATION:
9,350 ft. N. Cottonwood Creek

TRAIL ENDING ELEVATION:
8,900 ft. S. Cottonwood Creek

ACCESS:
#1. North Cottonwood Creek: From US Hwy 2 at north end of Buena Vista, take Co. Rd 350 to west about 2 miles to Co. Rd 261. Turn north for about 1 mile to Co. Rd 3,65 on left. Follow Rd. 365 about 3-1/2 miles to Trailhead. Road 365 is very rough and recommended for high clearance vehicles only.

#2. Middle Cottonwood Creek: From center of Buena Vista at traffic light take the Cottonwood Pass Rd (Co. Rd No. 306) west about 9 miles to the Avalanche Trailhead.

#3. South Cottonwood Creek: One half mile south of Co. Rd 306 about 7 miles west of Buena Vista on Co. Rd 344. Trail Crosses road.

ATTRACTIONS AND CONSIDERATIONS:
This segment of the trail offers exceptional scenery. The trail climbs to a high elevation overlooking the Arkansas River valley while passing through a variety of life zones. Between Silver Creek and Middle Cottonwood Creek the trail passes through the Collegiate Peaks Wilderness. Undeveloped campsites are found at either end of this segment or at Middle Cottonwood The trail is used by some for access to 14,196 ft elevation Mount Yale. This trail is for foot and horse useonly. Mechanized transportationincluding motorcycles, mountain bicycles and hang gliders are prohibited in the Wilderness.

DIFFICULTY: MORE DIFFICULT
DISTANCE: 8.0 MILES

TRAIL
1776F-G
COLORADO*

USE: Light
No motorized vehicles, bikes or hang gliders

RECOMMENDED SEASON:

| SPRING | **SUMMER** | **FALL** | WINTER |

* COLORADO TRAIL
N. COTTONWOOD TO S. COTTONWOOD CREEK

TRAIL BEGINS: (This Segment) North Cottonwood Creek at Chaffee Co. Rd 365.
Elev. 9,350 ft. (2856km)

TRAIL ENDS: (This Segment) South Cottonwood Creek at County Road No. 344.
Elev. 8,900 ft. (2719km)

TRAIL PROFILE (ONE-WAY):

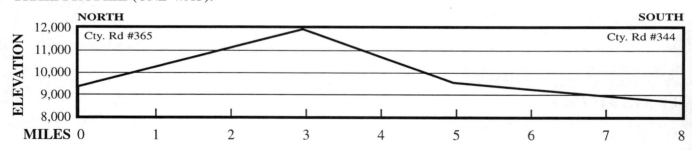

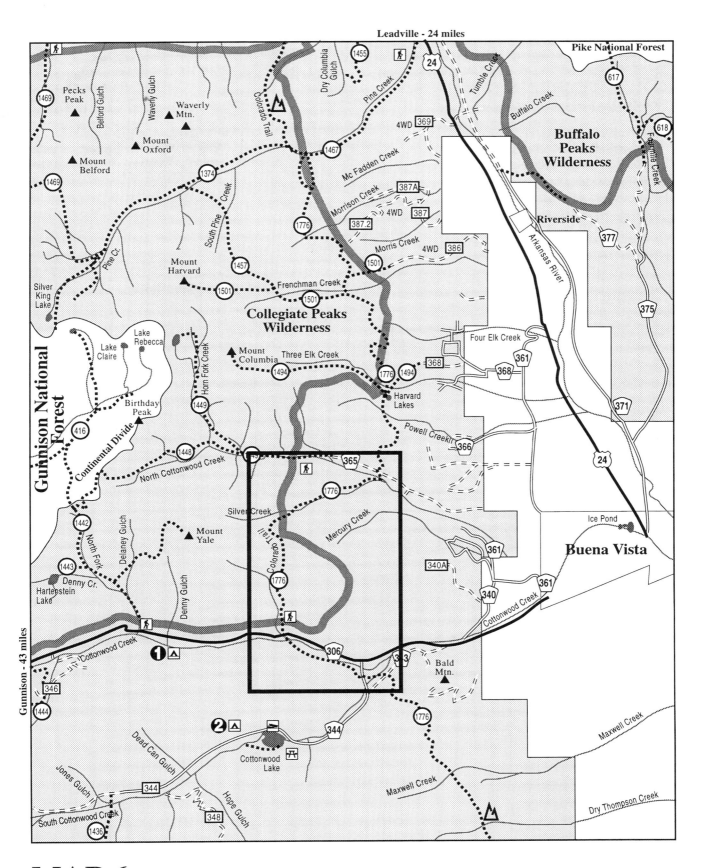

MAP 6
TRAIL #1776F-G

ROCKY MOUNTAIN REGION

MAP NUMBER(S): 6 and 9

NATIONAL FOREST: SAN ISABEL
RANGER DISTRICT: SALIDA

USGS: BUENA VISTA WEST, MOUNT ANTERO QUADS

DIFFICULTY: MODERATE

DISTANCE: 12.0 MILES

TRAIL 1776G-H COLORADO*

UPDATED: MARCH 1988

TRAIL BEGINNING ELEVATION:
8,900 ft. Cty. Rd #306

TRAIL ENDING ELEVATION:
8,170 ft. Chalk Creek

USE: Light
No motorized vehicles

RECOMMENDED SEASON:

SPRING	SUMMER	FALL	WINTER

ACCESS:

#1. South Cottonwood Creek: From the center of Buena Vista at the traffic light take the Cottonwood Pass Rd No. 306 about 7 miles west to South Cottonwood Rd No.344. Trail crosses the road about 1/2 mi. south.

#2. Chalk Creek: About 6-1/2 mi. west of Nathrop on Chalk Creek Road No. 162 at west end of County Rd 291 at Forest Service information sign.

SPECIAL NOTICE: Flash flooding may occur in gullies below the Chalk Cliffs. Do not attempt to cross these during heavy rain storms.

ATTRACTIONS AND CONSIDERATIONS:

The section from the Mt. Princeton Road to Chalk Creek is temporarily located following Public Roads through Private lands. Please stay on the roads following the trail signs and route markers. Please do not bother property owners with requests for services, etc. Camping is suggested at South Cottonwood Cr., Maxwell Cr., or Dry Creek. Also you may camp south of County Road 290 south of Chalk Creek.

Motor vehicles are not permitted on this segment of the trail except on public roads. Camping is not permitted through the private lands or along Chalk Creek. Please observe all posted signs and directions. Thanks for your help!

* **COLORADO TRAIL**
S. COTTONWOOD CREEK TO CHALK CREEK

TRAIL BEGINS: (This Segment) South Cottonwood Creek at Chaffee Co. Rd 344.
Elev. 8,900 ft. (2719m)

TRAIL ENDS: (This Segment) Chalk Creek.
Elev. 8,170 ft. (2496m)

TRAIL PROFILE (ONE-WAY):

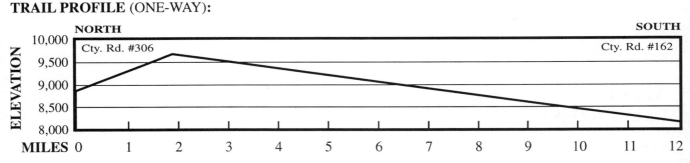

MAP 9 - South Half

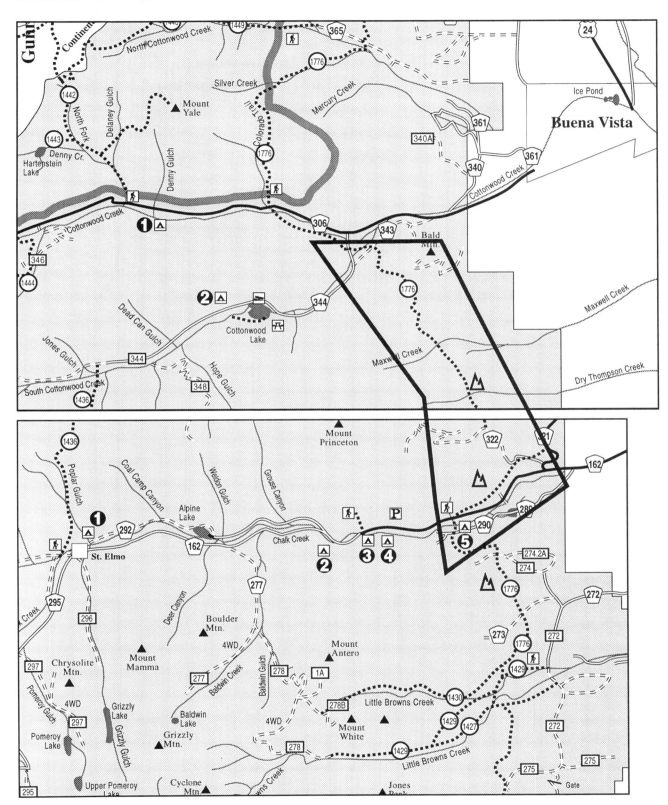

MAP 9 - North Half
TRAIL #1776G-H

© 1995 — *Outdoor Books & Maps, Inc.* • *Denver, Colorado* • (303) 629-6111

ROCKY MOUNTAIN REGION
MAP NUMBER(S): 9

NATIONAL FOREST: SAN ISABEL
RANGER DISTRICT: SALIDA

USGS: MOUNT ANTERO, MAYSVILLE QUADS

DIFFICULTY: MODERATE TO MORE DIFFICULT
DISTANCE: 19.5 MILES

TRAIL
1776 H-I
COLORADO*

UPDATED: MARCH 1988

TRAIL BEGINNING ELEVATION:
8,170 ft. Chalk Creek

TRAIL ENDING ELEVATION:
8,800 ft. US Highway 50

USE: Light

RECOMMENDED SEASON:

SPRING	SUMMER	FALL	WINTER

ACCESS:
#1. Chalk Creek: About 6-1/2 mi. west of Nathrop on Chalk Creek Rd No. 162 at Forest Service Information sign.

#2. Browns Creek: Trailhead 3 mi south of Nathrop then 2 miles, west on Co. Rd. 270 to Co. Rd. 272 then west and south about 4 mi to Browns Creek Trail. Then 1-1/2 mi west on trail 129.

#3. Blanks Cabin: From Poncha Springs go 2 miles west o US 50 to Co. Rd. 250 then north about 7 miles to end of road at Blanks Cabin Site. Cabin nonexist.

#4. Angel of Shavano: From Poncha Springs, go 6 miles west on US 50 to Maysville, then miles north on Co. Rd. 20 to Trailhead.

#5. U.S. Highway 50: About Nine miles west of Poncha Springs,. the trail crosses the highway.

ATTRACTIONS AND CONSIDERATIONS:
Side trips from this section of the trail feature 14,269 ft elevation Mount Antero by way of Browns Creek Trail No. 129, and to 1,229 ft elevation Mt Shavano and 14,155 ft elevation Mt Tabequache by way of Trail No. 1428. A good map and guide book should be consulted before attempting those side trips. Adequate time should be allowed.

Camping fees are charged at Mount Princeton and Angel Of Shavano C.G.'s

Camping is not permitted on Chalk Creek, except at Mt Princeton C.G. 1-1/2 mi to west. You may camp on the south side of Co. Rd. 290 south of Chalk Cr. South Hwy 50, there are no restrictions on camping. There are many undeveloped sites along the trail. Motorized vehicles are not permitted on this segment of the trail.

* COLORADO TRAIL
CHALK CREEK TO U.S. HWY 50

TRAIL BEGINS: (This Segment) Chalk Creek.
Elev. 8,170 ft. (296m)

TRAIL ENDS: (This Segment) U. S. Highway 50.
Elev. 8,800 ft. (2688m)

TRAIL PROFILE (ONE-WAY):

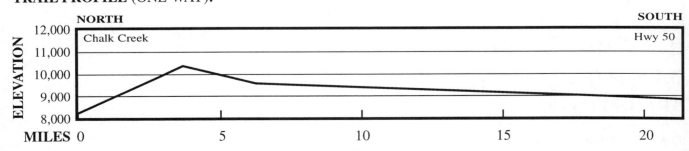

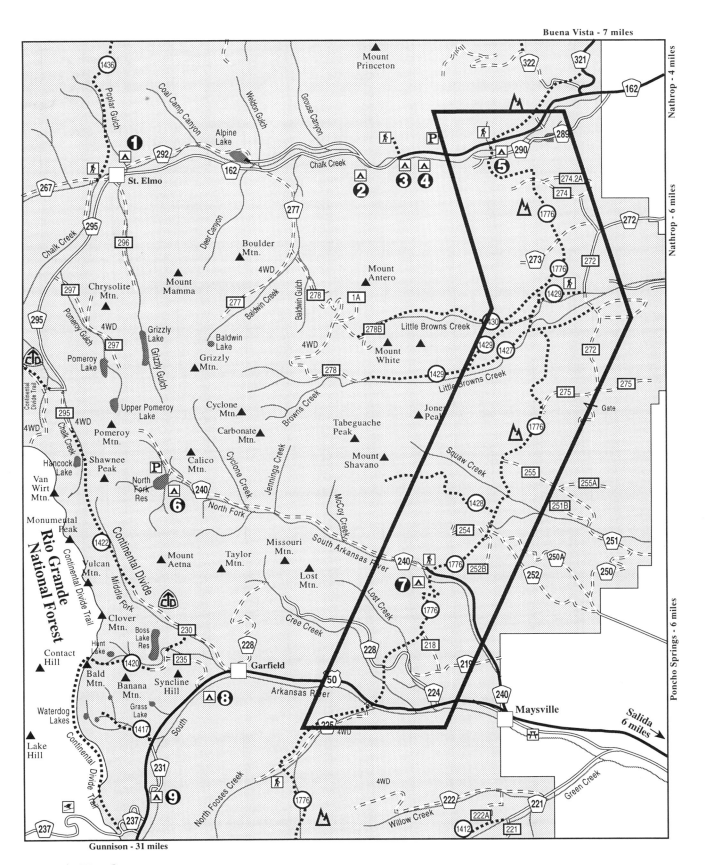

MAP 9
TRAIL #1776H-I

ROCKY MOUNTAIN REGION
MAP NUMBER(S): 9 and 12

NATIONAL FOREST: SAN ISABEL
RANGER DISTRICT: SALIDA

USGS: MAYSVILLE, PAHLONE PEAK, GARFIELD, MOUNT OURAY QUADS

DIFFICULTY: MODERATE

DISTANCE: 14.0 MILES

TRAIL 1776 I-J
COLORADO*

UPDATED: MARCH 1988

TRAIL BEGINNING ELEVATION:
8,800 ft. Highway 50

TRAIL ENDING ELEVATION:
10,840 ft. Marshall Pass

USE: Light
Limited motorized use

RECOMMENDED SEASON:

SPRING	SUMMER	FALL	WINTER

ACCESS:

#1. U.S. Highway 50: Trail crosses highway about 9 miles west of Poncha Springs at Fooses Creek road Co. Rd. No. 225. Park on south side of Hwy.

#2. South Fooses Creek: About 3 miles southwest of US Hwy 50 on Fooses Creek road (Co.Rd. 225). Trail follows road on this section.

#3. Marshall Pass: From Poncha Springs go 5 miles south on US Hwy 285 to Co. Rd 200, then west and south about 1 miles to summit of pass.

ATTRACTIONS AND CONSIDERATIONS:

From US Hwy 50, the trail follows the county road for about 3 miles before it begins to climb to the Continental Divide by way of South Fooses Creek. Camping is suitable along South Fooses Creek and at Marshall Pass. There is a primitive trail shelter along the Green Cr. Trail (No. 1412) a short distance off of the Colorado Trail. Marshall Pass is on the abandoned railroad grade of the Denver and Rio Grande R.R. route which once went to Gunnison. The high peaks along this segment are important Bighorn Sheep range.

Motorized vehicles are prohibited on the section from South Fooses to the top of the Divide. They are allowed on the section which is the Divide Trail which runs from Monarch Pass to Marshall Pass. The trail follows exposed ridges at high elevations for several miles. Hikers should be cautious of lightning storms and avoid adverse weather.

* COLORADO TRAIL
U.S. HWY 50 TO MARSHALL PASS

TRAIL BEGINS: (This Segment) U.S. Highway 50.
Elev. 8,800 ft. (2688m)

TRAIL ENDS: (This Segment) Marshall Pass
Elev. 10,840 ft. (3306m)

TRAIL PROFILE (ONE-WAY):

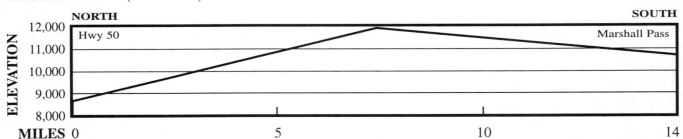

MAP 9 - South Half

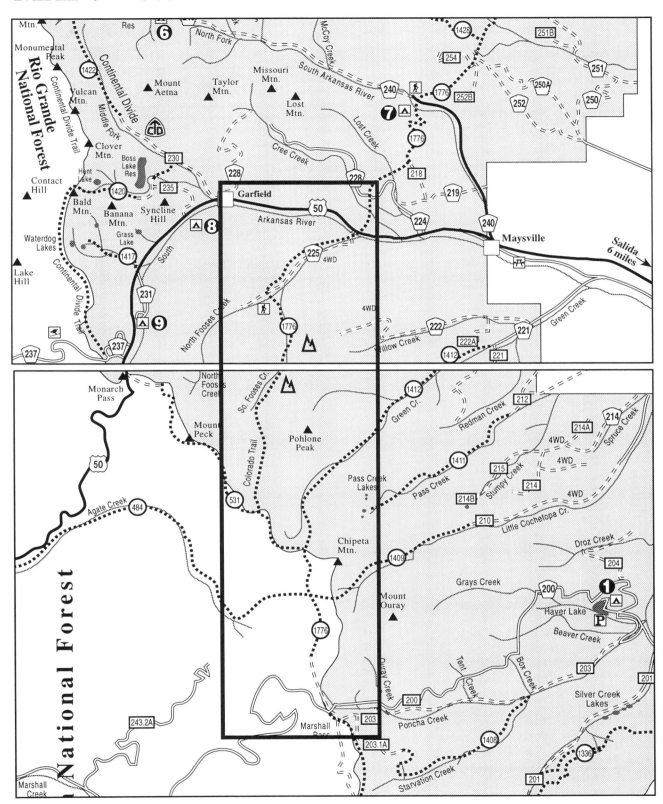

MAP 12 - North Half
TRAIL #1776I-J

© *1995 — Outdoor Books & Maps, Inc.* • *Denver, Colorado* • *(303) 629-6111*

ROCKY MOUNTAIN REGION
MAP NUMBER(S): 12

NATIONAL FOREST: SAN ISABEL
RANGER DISTRICT: SALIDA

USGS: MOUNT OURAY, CHESTER, BONANZA, SARGENTS MESA

UPDATED: MARCH 1988

TRAIL BEGINNING ELEVATION:
10,840 ft.

TRAIL ENDING ELEVATION:
11,050 ft.

ACCESS:
#1. Marshall Pass: From Poncha Springs go 5 miles south on US Hwy 25 to Co. Rd 200, then west and south about 1 mile to summit of Marshall Pass.

#2. Sargents Mesa: From Saguache, take State Hwy 114 about 11 miles west to the Jacks Creek Rd (FDR 853), then northwest on this road about 6 miles to the Houghland Gulch Road (rDR 850), then northwest and north on this road about 7 miles to the trail.

ATTRACTIONS AND CONSIDERATIONS:
From Marshall Pass south the trail follows Gunnison N.F. Forest Road Number 243.2D, a logging road closed to public vehicles for about 2 miles as it climbs to the Divide and crosses to the east side on the San Isabel N.F. for another 2 miles. Then it turns to the west following the Continental Divide on tne trail also identified as trail no.486. This is also the route of the Continental Divide National Scenic Trail corridor which will be identified and marked within a few years.

Much of this segment of the trail follow the Continental Divide resulting in long sections without access to water. From Marshall Pass, It is about 10-1/2 rniles to water at Tank Seven Creek. From there it is another 12 or more miles to the next water at Upper Razor Creek on the next segment south.

DIFFICULTY: MODERATE TO DIFFICULT
DISTANCE: 16.5 MILES

TRAIL
1776 J-K
COLORADO*

USE: Light

RECOMMENDED SEASON:

SPRING	SUMMER	FALL	WINTER

*** COLORADO TRAIL**
MARSHALL PASS TO SARGENTS MESA

TRAIL BEGINS: (This Segment) Marshall Pass
Elev. 10,840 ft. (3306m)

TRAIL ENDS: (This Segment) West end of Sargents Mesa at Hougland Gulch Rd (Rio Grande N.F. FDR 850)
Elev. 11,050 ft. (3425rn)

SPECIAL NOTE: See Gunnison and Rio Grand National Forests for additional ROG sheets as trail continues to south.

TRAIL PROFILE (ONE-WAY):

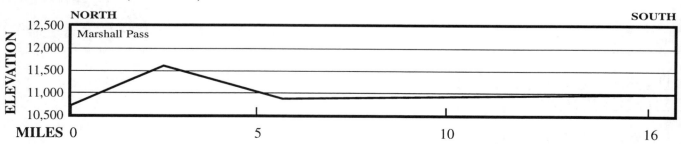

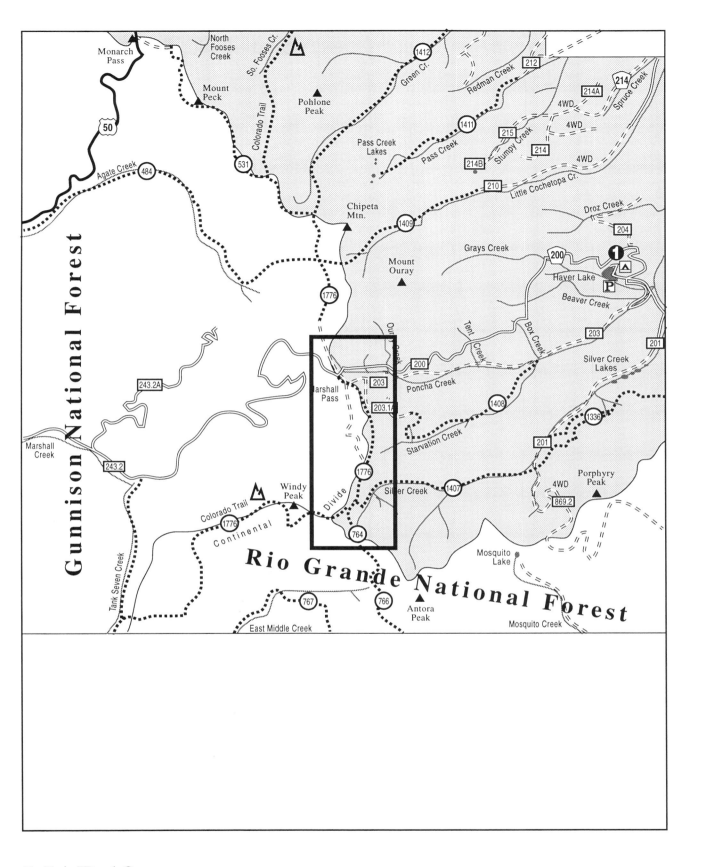

MAP 12
TRAIL #1776J-K

TURQUOISE LAKE

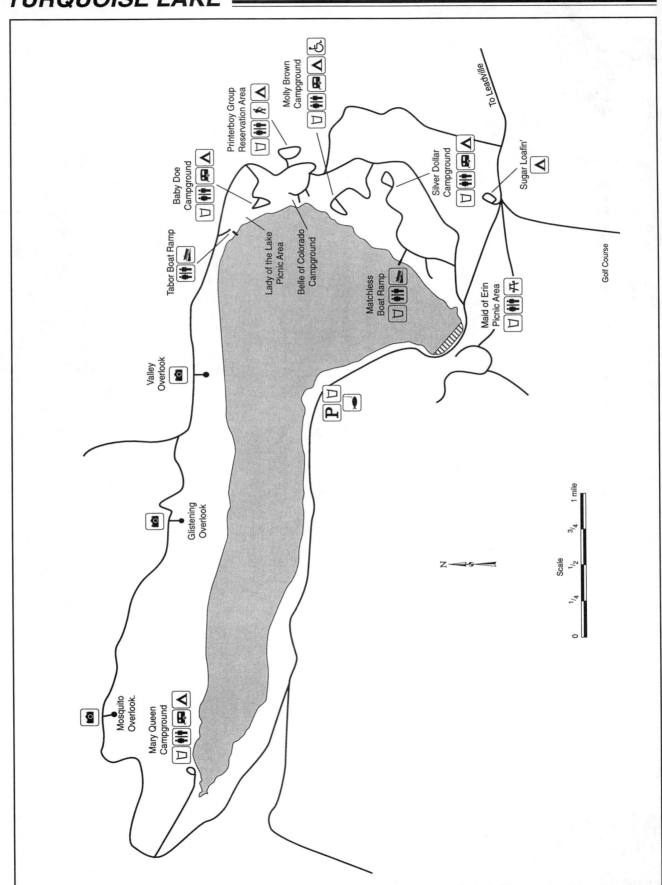

TURQUOISE LAKE

DIRECTIONS: From Leadville, west on the 4 1/2 mile, paved Turquoise Lake Road or a marked turnoff south of Leadville on U. S. 24

FEE: Camping fees.

SIZE: 1,650 surface acres when full.

ELEVATION: 9,869 feet

MAXIMUM DEPTH: 80-90 feet

FACILITIES: Full Service campgrounds

BOAT RAMP: Two concrete boat ramps are available. The Tabor Boat Ramp is on the northeast end of the lake; the Matchless Boat Ramp is on the southeast end.

RECREATION: Fishing, boating.

CAMPING: Campgrounds generally are open from Memorial Day weekend until Labor Day weekend. Seven campgrounds available. On the west end of the lake, the May Queen campgrounds offer 34 campsites. On the east end of the lake, Baby Doe offers 50 sites, Molly Brown 49, Belle of Colorado 19 (tent-only walk-in), Father Dyer 36, Tabor 20 and Silver Dollar 45. Most of the campgrounds feature toilets, picnic tables and fire grates. Water also is available. Several picnic areas are located around Turquoise Lake. Unreserved campsites are handled on a first-come, first-served basis. A group campsite, Printer Boy, is available. It has a capacity of 180 people and may be reserved.

FISH: Snake River cutthroat and rainbow, brown and mackinaw trout. Fishing from shore, use salmon eggs, cheesebait and Power Bait on the bottom.

CONTROLLING AGENCY: San Isabel National forest.

INFORMATION: Leadville Ranger District
(719) 486-0749.

MAP: San Isabel National Forest map.

Turquoise Lake is located five miles west of Leadville. The lake is man-made and is a part of the Lake Fork drainage in the upper Arkansas Valley. It has an approximate elevation of 9,900 feet in mountainous terrain, and is backed against the Sawatch Mountain Range, which includes Colorado's two highest peaks, Mount Elbert and Blunt Massive.

Turquoise Lake is incorporated into the Fryingpan-Arkansas Water Project. This project is diverting water from the west slope of the continental divide to the more populated eastern slope. Water is diverted by a system of tunnels and at Turquoise Lake there are two tunnels, the Charles Boustead Tunnel and the Homestake Tunnel. Turquoise Lake is one of several storage areas in the Fryingpan-Arkansas Project.

Leadville, as well as the Turquoise Lake area, is rich in a colorful history. Leadville's glamorous past began during the "Gold Rush" in California Gulch, southeast of Leadville, in 1869. Many of Leadville's prominent personalities and places of this era are now honored as recreation sites along Turquoise Lake.

In 1860, after leaving the 59'ers in the Pikes Peak region one expedition crossed the Mosquito Range and traveled up the Arkansas River into what is now called California Gulch. There, one man panned the first valuable pay-dirt. His name was Abe Lee. A fisherman parking area is now honored with his name.

Some five years after the 1869 gold rush into California Gulch, the gold and the population started to dwindle. Employing new methods of underground "hardrock" mining, one mine began to pay off after several years. The mine was Printer Boy. Now Printer Boy campground at Turquoise Lake carries this name.

In 1874, Alvinus B. Wood and William H. Stevens analyzed some rock and sand which had been clogging prospector's mining equipment. They found it to be lead carbonate, richly lined with silver. The silver boon was on! Leadville grew from 300 to over 30,000 residents in less than two years. Mining claims and mines sprang up, seemingly overnight. Four of these mines are now names of recreation sites in the Turquoise Lake Recreation Area. They are Lady of the Lake picnic area, Maid of Erin picnic area, Belle of Colorado campground, and May Queen campground.

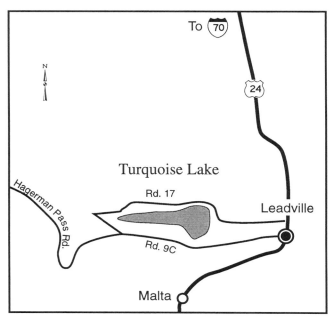

TWIN LAKES

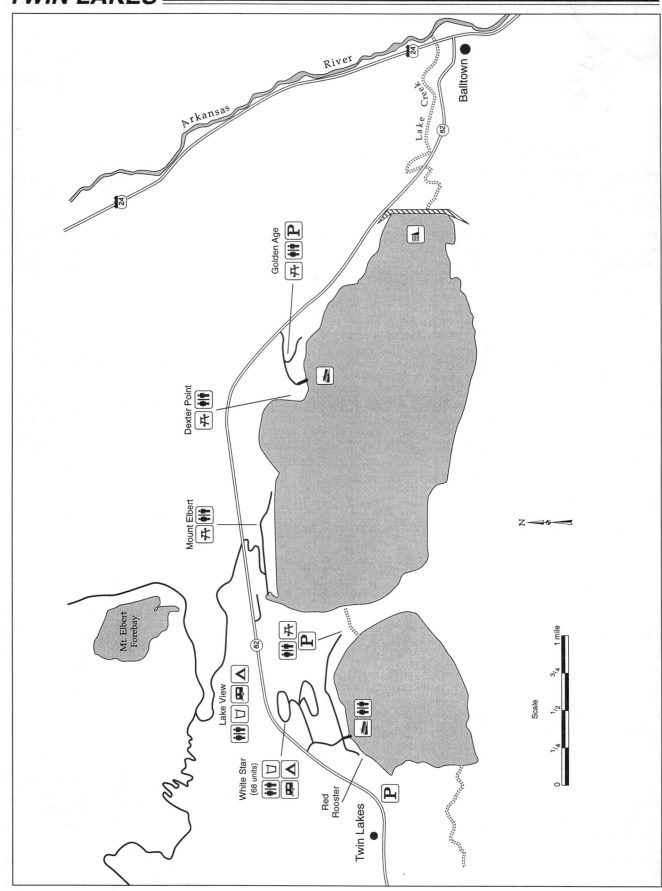

TWIN LAKES

DIRECTIONS: From Colorado Springs, west on U. S. 24 to Buena Vista. Follow U. S. 24 north out of Buena Vista for 17 miles to Colorado 82. Six miles west on 82 to the Twin Lakes Reservoirs.

FEE: Camping fee.

SIZE: 1,700 surface acres of water.

ELEVATION: 9,700 feet

MAXIMUM DEPTH: 70 feet.

FACILITIES: Full service campgrounds.

BOAT RAMP: Two ramps, both paved.

RECREATION: Fishing, boating.

CAMPING: Five campgrounds, 216 total campsites. Fire pits, picnic tables, vault toilets, drinking water all available. Among the most scenic reservoirs in Colorado, located at the base of the Continental Divide.

FISH: Lake trout, rainbow, brown and kokanee. Lake trout are caught trolling with sucker meat, Rapala and Flatfish lures. Rainbow are caught from shore using salmon eggs and Power Bait. Brown are caught trolling with nightcrawlers on Pop Gear or Rapala lures. Kokanee are caught trolling with Redmagics and Kokanee Killers. Lake trout limit is one fish. All lake trout between 22-34 inches must be returned to water immediately.

CONTROLLING AGENCY: San Isabel National Forest.

INFORMATION: Leadville Ranger Districtt
(719) 486-0749

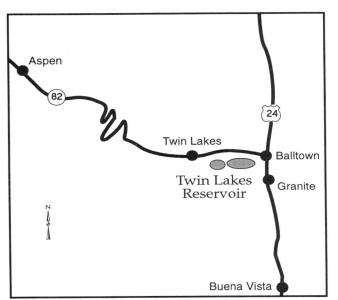

Twin Lakes are natural glacier-formed lakes which have been enlarged to provide additional storage capacity for the Fryingpan/Arkansas Project of the U.S. Department of Interior, Water and Power Resources Service. Water from the Fryingpan River drainage on the western slope of the Continental Divide is brought through the Boustead Tunnel to Turquoise Lake. Then it is transported by conduit to the forebay north shore of Twin Lakes.

During periods of low power use, the turbines are reversed to pump water back into the forebay. Here it is held until the demand for electric power increases. The water is again cycled through the turbines.

Fishing and boating are favorite activities at Twin Lakes. The Lakes are especially noted for their mackinaw trout with occasional trophy sized "macks" up to 35 inches in length being taken by anglers. State laws apply to boating and fishing.

The Colorado Trail north from Twin Lakes offers hiking to Tennessee Pass over 30 miles to the north and south of Twin Lakes to Cottonwood Creek west of Buena Vista 25 miles to the south. The Colorado trail also passes through the recreation area connecting with auxiliary trails to Willis Lake, Mt. Elbert and Interlaken.

Historic Interlaken, turn-of-the-century resort complex, is being restored on the southwest shore of the lower lake. Both Interlaken and Twin Lakes Village have been placed on the National Register of Historic Places. While restoration work is in progress, visitors will not be permitted.

The area around Twin Lakes is closed to camping except in designated areas. Users should note and read other special regulations applicable to the Twin Lakes area which are posted throughout the recreation area.

Twin Lakes Recreation Area consists of one campground, two boat ramps, five fishermen parking areas and one picnic area. The majority of the developed recreation areas are located on the north side of Twin Lakes, approximately 20 miles south of Leadville.

SPECIAL RESTRICTIONS:
Keep fires inside stoves, grills, or fireplace rings provided at picnic sites. Do not deface, remove, or destroy plants and trees. Pets must be on a leash not longer than six feet. Drive all motor vehicles only on developed roads within the area. The area is closed to use of firearms or fireworks.

COLORADO FISHING RECORDS

Colorado Division of Wildlife

Bluegill   Colorado Record - 2 lbs., 4 oz.
This sunfish has a short and deep body. As with all sunfish, the dorsal (top) fin is not split. The bluegill has a small mouth on a short head and a dark gill flap with no trim. There are parallel vertical bars on the side with long, pointed pectoral (side) fins. A male bluegill in breeding colors has brilliant blue fins and red-orange stomach. The female bluegill is dark on the back with vertical stripes on the body. Bluegills are best caught in the morning or evening using small tackle ranging from cork and worm to delicate dry flies. Once one bluegill is located, others will be nearby. Bluegill spawn in colonies from late spring to August, building nests on gravel, sand, mud, leaves, or sticks in 1-4 feet of water. As summer heat becomes extreme, these fish move to deeper water and the shade of weed beds.

Green Sunfish   Colorado's Record - 1 lb., 3 oz.
This fish is similar in appearance to the bluegill, but has larger mouth and is olive in color with short, rounded pectoral fins and yellow trim on the fins. This stocky fish is found in both streams and impoundments and spawns in shallow areas from June to mid-August. Like most sunfish, this spotty panfish can be taken with crickets, worms and other bait rigged under a bobber or with small lures jigs and flies.

Pumkinseed
Another similar fish, the pumpkinseed has a red-orange dot on its gill flap with trim and irregular clusters of orange spots on its body. The pumpkinseed prefers weed patches, docks and sunken vegetation. It stays close to shore and is easily taken by a variety of baits ranging from juicy grubs or worms to small lures or wet flies.

White Crappie   Colorado's Record - 4 lbs., 3 and three-quarter oz.
This sunfish has a flat, short, and deep body with an unsplit dorsal fin, a large mouth and a longer head than other sunfish. Its body is silver-white speckled with small irregular black blotches. White crappie have fewer than seven dorsal fin spines and are tolerant of warm, muddy water. These fish congregate in large schools during the spring, when small jigs fished around submerged brush piles work well. Crappie are often caught on small minnows, jigs, poppers, plugs and spinners. Late March and May are the best times to fish for crappie in the metropolitan area.

Black Crappie   Colorado's Record - 3 lbs., 4 oz.
Closely related to the white crappie, this fish has from seven to nine spines on its dorsal fin and prefers clear, weed-covered reservoirs.

Largemouth bass   Colorado's Record - 10 lbs., 6 and one-quarter oz.
This sunfish has an elongated body with the dorsal (top) fin in two lobes. The body is silvery in color with brown on top and a dark, blotchy horizontal band. When the fish's mouth is closed, its jaw extends beyond its eye. The largemouth bass spawns in 18-36 inches of water in late spring. It can be caught on a number of natural baits, including frogs, crayfish, worms, grasshoppers and minnows. Artificial baits such as flies, poppers, plugs, artificial worms and spoons are effective. Dawn and dusk are good times to fish for bass around submerged brush or underwater drop-off.

Smallmouth Bass   Colorado's Record - 5 lbs., 8 oz.
Similar in appearance to the largemouth bass, this fish is distinguished by greenish sides with vertical bars; its jaw also extends only to the middle of its eye when the fish's mouth is closed. Favorite baits are minnows, crayfish, surface and underwater lures, poppers and flies fished around brush, weeds and underwater structures in early morning and evening.

Walleye   Colorado's Record - 16 lbs., 9 oz.
This fish has an elongated body with the dorsal (top) fin completely divided and a large mouth in a slender head with several fang-like teeth. Color is whitish on the bottom to yellowish on the sides with dark, irregular blotches; the lower lobe of the caudal (tail) fin has a white blotch. Walleye can be caught on most artificial lures with deep-running spoons, plugs, worm or minnow rigs, or spinnerfly combinations working well. Large jigs, crank baits, and spinners over shallow rocky areas catch walleye in the spring.

Yellow Perch   Colorado's Record - 2 lbs., 5 oz.
Closely related to the walleye, this fish has an elongated, yellowish body with six or seven dark, vertical bars. Its dorsal fin is completely divided, the lower fins are often trimmed in yellow-orange and the end of the gill flap has a sharp point. It lacks the fang-like teeth of the walleye. Probably Colorado's most abundant game fish, yellow perch bite best at midday and toward evening. Small flies and spinners or natural bait, such as worms or grubs fished a foot or two off the bottom, work well summer and winter (many people fish for perch through the ice).

Channel Catfish   Colorado's Record - 33 lbs., 4 oz.
These native fish have spines in the dorsal and pectoral fins, long barbels under the mouth and adipose fin. The body is scaleless with a light to silvery lower part and dark upper body color. Younger channel catfish have a few dark spots on their bodies. All have a forked tail. Channels spawn in early summer when water temperatures reach 70 to 80 degrees F. Night fishing, using a variety of bottom bait - including night crawlers, minnows, crayfish, chicken innards and flavored dough balls - provide the best action. Many fisherman let this wary fish run for several seconds on an open bail before setting the hook.

Black Bullhead   Colorado's Record - 4 lbs., 10 oz.
This small catfish is also native to Colorado. It is distinguished from the channel catfish by a yellow or brownish color; in addition, its tail is squared, not forked. Chin barbels are dark, pectoral fins are smooth. These fish will take worms, grasshoppers and minnows fished on the bottom. Evening fishing is best, but bullheads will also bite in daylight hours.

Northern Pike   Colorado's Record - 30 lbs., 10 oz.
This fish has a dorsal fin set far back on a long, greenish-gray body with whitish, irregular spots. The head is duckbill shaped with many sharp teeth. Spawning over underwater vegetation immediately after ice-off, this fierce-looking predator provides exciting fishing. Large, shallow-running spoons and active live bait along with heavier tackle are a must to land a hard-fighting, 40 inch northern. Casting flashy spinners, spoons and lures over weedy areas against the shore or trolling shallow plugs or live bait often works for these fish. Wire leaders should be used to prevent the sharp teeth from cutting the line.

Tiger Muskie   Colorado's Record - 27 lbs., 3 oz.
This fish is similar to the northern, but has a lighter colored body with dark vertical side bars (tiger stripes). Statewide bag and possession limits 1 fish, 30 inches or longer.

Common Carp   Colorado's Record - 27 lbs., 4 oz.
Carp have two small barbels on each side of the jaw, a large serrated spine in a long, single dorsal and anal fin and large scales. The three varieties of common carp found in Colorado are scaled, mirror (partially scaled) and leather (scaleless). Carp thrive in warm, shallow water with plenty of aquatic vegetation. They spawn in late May and into June. Fish for carp on the bottom with bait; flavored dough balls made of bread or cornmeal work best. Grass carp are also found in the metro area, where they have sometimes been stocked to help control vegetation. These carp do not have barbels and do not have spines in the anal or dorsal fin.

Rainbow Trout   Colorado's Record - 18 lbs., 5 and on-quarter oz.
This trout has very fine scales, an adipose fin, a silver body with small spots speckling the side and often a horizontal pink streak. This fish is the mainstay of the Division of Wildlife's hatchery system because it is relatively easy to raise to catchable size. Famed for their fighting abilities, rainbows can best be caught during their prime feeding periods - early morning and late evening. They can be caught on a variety of baits, lures and flies.

Cutthroat Trout   Colorado's Record - 16 lbs.
Colorado's only native trout species, the cutthroat (also called native) trout has a brilliant crimson slash mark on each side of the throat beneath the lower jaws. Many cutthroat now found in Colorado are hybrids which are the product of past stocking programs where different varieties were introduced into the state. All cutthroat varieties native to Colorado have very few spot on the front part of their body below the lateral line. These trout are mostly found in remote mountain streams and lakes, which hatchery-reared finger-lings are often stocked. They can be caught with a variety of baits - such as red fish eggs and water insects - spinning lures and wet or dry flies.

Brook Trout   Colorado's Record - 7 lbs., 10 oz.
These trout have light, wormy streaks on dark bodies. Pectoral, pelvic and anal fins often orange, edged with black and white. Brook trout thrive in cold, high elevation streams and feed on aquatic and terrestrial insects. They will rise readily to a wide range of small lures, baits and flies.

Brown Trout   Colorado's Record - 30 lbs., 8 oz.
These trout are distinguished by large, dark spots. The spotting pattern is composed of a combination of black spots and red-orange spots surrounded by blue. The brown trout is found at lower elevations because it can withstand warmer water temperatures more than other trout. Browns are hard to catch, but streamers or small dry flies fished in late summer evenings can produce in streams. Fall fishing with large flies and lures at lake inlets at dawn can land a lunker.

SECTION II

BACKPACKING

Backpacking Is Freedom

Backpacking offers freedom to the forest traveler. You have no worries, other than your own. You become part of a scenic landscape and survive in a primitive environment with few modem conveniences. Self-sufficient, yes, but with this freedom goes an individual responsibility to care for the environment and respect the rights of those you meet along the way and those who follow you.

Backpacking is not limited to supermen and superwomen. However, it does require physical stamina and a genuine liking for the isolation in the remote country. Overnight backpacking trips should be undertaken only by those who have hiked easier mountain or forest trails and are familiar with back-packing techniques.

Leave No Trace

For thousands of years our wildlands have existed in a complex ecological interrelationship. This interrelationship can be easily upset or even destroyed. Once damaged, some plants and soils may not recover in our lifetime. Today, nature is struggling in many backcountry areas to cope with results of unacceptable backpacking, overnight camping techniques, and heavy use.

Unappreciative or uninformed backpackers who have no enthusiasm for preserving the land are now in the minority. Even so, many backcountry areas are "camped out." Firewood is scarce or nonexistent. Unnatural fire blackened rocks and fire scars dot the landscape, and small green trees and ground cover are gone. In many areas, the streams are no longer safe for drinking. Several groups of people camping around the same lake lower the quality of the "backcountry experience" through noise and visual pollution.

Laws and regulations are being enforced to correct and eliminate these situations, but cooperation, proper attitudes, and voluntary actions of visitors are better ways to preserve the land.

The concept of taking only pictures and leaving only footprints evolves from backpacker awareness.

Special regulations

Permits are required in some areas of the backcountry. Permits can be obtained from the local offices of the land managing agency. The permit must be obtained in advance and must be in your possession during your visit.

Group Size

In many backcountry areas the maximum number of people in a group is restricted. Large groups are destructive. Check to determine allowable group size.

Trail Courtesy

When hiking it's quite possible you may encounter trail riders along with pack stock. Livestock are easily spooked from unseen sources, it is best to make your presence known. When stock approaches, step off on the lower side of the trail while the stock passes. Be courteous in sharing the trail with others

Fishing and Hunting

Write in advance to the State Game and Fish Department for fishing and hunting rules and licenses.

Fishing and hunting are authorized under State regulations. Check with the local Game and Fish official before entering areas to fish or hunt because regulations vary.

In every jurisdiction, the "plinking" gun used to destroy chipmunks, song birds, and other wildlife is held in contempt, and it is usually illegal.

Pets

Regulations differ on taking pets into the backcountry so check with the local Ranger regarding restrictions. Remember: dogs and cats are predators by nature and will instinctively chase forest birds and animals; horses and dogs don't mix, so physical restraint of the dog is necessary; and bears and dogs don't mix.

You know your pet but other persons do not. Many areas have leash restrictions, especially on or within specified distances (usually 300 feet) of well-traveled trails or in heavily used areas. Show respect for other persons and wildlife by keeping your pet under physical restraint or better yet, you might consider leaving your pet at home.

Awareness and Techniques

Backpacker awareness means understanding how you fit into the backcountry scene and not leaving evidence of your visit. If such awareness were practiced, all visitors would have the same opportunity to experience the natural scene. This awareness is intended to create backpacker recognition of the fragility of backcountry areas and a personal commitment to the care and wise use of this land.

If we could look back at the Rockies, the Southwest, or the Lake States in 1830, we would see a land devoid of cities, roads, and vehicles, inhabited only by Indians and mountain men. When traveling the backcountry, the mountain man's priorities were adventure, monetary gain, and personal survival. Today's visitors to the backcountry seek solitude, primitive recreation, and natural scenery.

Yesterday's mountain man left no sign of his presence in Indian country. Today, backpackers should leave no signs of their presence so that the next person can enjoy a natural scene and the solitude it portrays. You must tread lightly so nature can endure and replenish.

Trip Planning

The first step of awareness and backpacking technique is planning your trip. As one of numerous visitors in the backcountry, plan your trip carefully to protect yourself as well as the environment.

For a carefully planned trip, consider:

Maps to plan access, take-off and return points, route of travel, approximate camping areas, and points of attraction to visit;

Proper lightweight equipment to safely cope with the elements and your recreational pursuits;

Food for the entire trip, packed in lightweight containers such as plastic bags;

Number of persons in the party and their abilities; and

Regulations and restrictions that may be applicable.

Experience will help you refine planning skills, equipment, and techniques. However, evenings at home with how-to-do-it books, practice in putting up tents or shelters from groundcloths, and trying out dehydrated foods or home recipes will spark the imagination and eliminate some mistakes.

What You Need For:

Camping: Pack a tent or tarp for a shelter, sleeping

bag, foam pad, lightweight stove, cooking utensils, dishes and cutlery, and a small flashlight with extra batteries and bulb. Food should include snacks for the trail.

Clothing: Bring slacks or jeans- 2 pairs, long-sleeved cotton shirts, at least 2 wool shirts or a sweater, parka windbreaker, wool socks -2 changes, underwear, camp shoes and socks, rain gear, rain shirt, poncho or nylon raincoat, and handkerchiefs.

First Aid Kit: (you can make your own): Bring adhesive bandages, compresses, 4-inch elastic bandages, triangular bandage, antiseptic, aspirin, eye wash, adhesive tape, insect repellent, sunscreen lotion, mole-skin for blisters, tweezers, and chapped lip medication.

Hiking: Wear footwear with eyelet's and lacing. Most backpackers prefer 6- to 10-inch laced boots with rubber or synthetic soles. Footwear should be "broken in" and fit comfortably over two pairs of socks, one light and one heavy. Take extra laces. Pack a pair of soft sole shoes. After a day of hiking, they will feel comfortable as well as being less damaging to campsite vegetation.

Personal Sanitation: Carry a lightweight shovel or trowel, and toilet paper.

Extra Comfort: Bring dark glasses, rope (nylon cord), knife, small pliers, waterproof matches, biodegradable soap, a towel, needle, thread, and safety pins.

When to Travel

Time your trip according to climatic conditions. For example, in California's Hoover Wilderness backpacking season is about 2 months long -July and August. Even then, the hiker and camper should be prepared for all kinds of weather including rain, summer blizzards, extreme cold, and heavy winds. In the Colorado mountains, conditions are usually favorable for traveling June 15 to October 1, but in the Northern Rockies, the best time for a trip is between July 15 and September 15. If you go into the high country too early, snow may interfere with travel, streams tend to be high and difficult to cross, fishing may be poor, and meadows and trails are apt to be soft and subject to damage. July and August are subject to intense afternoon thunder and lightning storms in the alpine areas. August and early September often provide the best weather for travel in the high country, With little bother from insects.

If you choose a route without trails do not mark the trees, build rock piles, or leave messages in the dirt. A group should spread out rather than walk one behind the other (especially in tundra or meadow areas). Ten people tramping in a row can crush plant tissue beyond recovery and create channels for erosion.

Hike in groups of 4 to 6 people at most; 4 is the best number, especially during off-trail travel. In case of sickness or injury, one person can stay with the victim while two people go for help. Use your judgment in breaking your group into smaller units to reduce visual impact and to increase individual enjoyment and self-reliance.

Pick up any litter along the route; have one pocket of your pack available for trash.

Avoid removing items of interest (rocks, flowers, wood or antlers). Leave these in their natural state for others to see.

Allow horses plenty of room on trails. Horses may be frightened by backpack equipment. It is best to move off the trail. Everyone in your group should stand off to the downhill side of the trail. Avoid sudden movements as horses pass.

Help preserve America's cultural heritage by leaving archeological and historical remains undisturbed, encourage others to do the same, and report your discoveries to the local ranger.

Locating a Campsite

Check with the local ranger for suitable camping areas; then plan your trip to avoid areas that need to recover from overuse.

If other parties are close to where you want to camp, move on or choose your campsite so that terrain features insure privacy. Trees, shrubs, or small hills will reduce noise substantially. Out of respect for nearby campers keep the noise level low at your campsite.

Use an existing campsite whenever possible, in order to reduce human impact. If selecting a new campsite, choose a site on sandy terrain or the forest floor rather than the lush but delicate plant life of meadows, stream banks, fragile alpine tundra, and other areas that can be easily trampled or scarred by a campfire.

Camp at least 200 feet away from water sources, trails, and "beauty spots" to prevent water and visual pollution.

Take a little extra time to seek out a more secluded area. It will increase your privacy and that of other visitors.

Arrange the tents throughout the campsite to avoid concentrating activities in the cooking area.

Avoid trenching around your tent, cutting live branches or pulling up plants to make a park-like campsite. If you do end up clearing the area of twigs, or pine cones, scatter these items back over the campsite before you leave.

A backcountry campsite should be reasonably organized. If you have laundry to dry or equipment to air out, try to make sure these items are not in sight of other campers or hikers.

Leave the area as you found it, or in even better condition.

Travel Light

Experienced backpackers pride themselves on being able to travel light. Rugged, sure footed backpackers will seriously explain that they cut towels in half and saw the handles off toothbrushes to save ounces. They measure out just the right amount of food needed and put it in plastic bags, which are light. They carry scouring pads with built-in soap, to eliminate dish soap and a dishcloth.

How much should you carry? It all depends on your physical condition and experience, the terrain to be covered, the length of the trip, and the time of year. The average is 30 pounds for women (maximum 35) and 40 pounds for men (maximum 50).

When figuring weight, count all items - the cup on your belt, the camera around your neck, the keys in your pockets.

Backcountry Travel

Travel quietly in the backcountry, avoid clanging cups, yells, and screams. Noise pollution lessens the chance of seeing wildlife and is objectionable to others seeking solitude. However, in "grizzly country" noises may keep the bears away.

Wear "earth colors" to lessen your visual impact, especially if you are traveling in a group. However, during hunting season a blaze orange hat and vest are advisable for your personal safety.

When tracking wildlife for a photograph or a closer look, stay downwind, avoid sudden motions, and never chase or charge any animal. Respect the needs of birds and animals for undisturbed territory. Some birds and small animals may be quite curious, but resist the temptation to feed them. Feeding wildlife can upset the natural balance of their food chain - your leftovers may carry bacteria harmful to them.

Stay on the designated path when hiking existing trails. Shortcutting a switchback or avoiding a muddy trail by walking in the grass causes unnecessary erosion and unsightly multiple paths. In the spring, travel across snow and rocks as much as possible; high mountain plants and soil are especially susceptible to damage during a thaw.

If you choose a route without trails, do lot mark the trees, build rock piles, or leave messages in the dirt. A group should spread

Hike in groups of 4 to 6 people at most; 4 is the best number, especially during off-trail travel. In case of sickness or injury, one person can stay with the victim while two people go for help. Use your judgment in breaking your group into smaller units to reduce visual impact and to increase individual enjoyment and self-reliance.

Pick up any litter along the route; have one pocket of your pack available for trash.

Avoid removing items of interest (rocks, flowers, wood or antlers). Leave these in their natural state for others to see.

Allow horses plenty of room on trails. Horses may be frightened by backpack equipment. It is best to move off the trail. Everyone in your group should stand off to the downhill side of the trail. Avoid sudden movements as horses pass.

Help preserve America's cultural heritage by leaving archeological and historical remains undisturbed, encourage others to do the same, and report your discoveries to the local ranger.

Campfires and Stoves

The mountaineer's decision to have a campfire was frequently influenced by the friendliness of the Indians. Today, your most important consideration should be the potential damage to the environment. A stove leaves no trace.

You should use a campfire infrequently and only when there is abundant dead wood available on the ground. Be very critical about the necessity for campfires. In many areas, wood is being used faster than it grows. In over camped areas or near timberline, choose an alternate campsite or use a portable stove.

In all areas fires should be completely out before you abandon the campsite. In some areas campfires are prohibited by regulations. Check with the public land management agency for local regulations.

If you do build a campfire remember:

All fires must be attended. Be aware of overuse. If your firepit is full of wood ash or our cooking area unnecessarily trampled, move your campsite to lessen the camping scar.

Fires should be built away from tents, trees, branches, and underground root systems.

Campfires should never be built on top of the forest floor. If there is a ground cover of needles and decomposed matter be sure to dig through it to the soil.

Be sure the firepit is large enough to prevent the possibility of the fire spreading.

Do not build fires on windy days when sparks might be dangerous, especially when the countryside is dry.

Types of Fires

If you come upon a fire ring in the backcountry and the surrounding area has not been over camped, make use of it. However, fires should not be ringed with rocks as this permanently blackens them. When there is no existing fire ring, use one of the following three types of fires to assure little impact.

Flat Rock Method: Spread several inches of carefully gathered bare soil on top of a flat rock over an area slightly larger than the fire will occupy, then build your fire as usual. Burn all wood completely. After the fire is out, crush and scatter any coals. After the soil is removed and the rock rinsed, the area will be virtually unscarred.

Pit Method: Remove sod or topsoil in several large chunks from a rectangular area, about 12"x24" (sufficient for a party of two). When excavating the pit, place the topsoil or sod neatly in a pile nearby, and the pile of bare soil around the firepit area to avoid drying out surrounding vegetation. If bare soil is not placed on top of sod surrounding the firepit, then the sod should be kept moist. On breaking camp, both the bottom and sides of the firepit should be cold to the touch. Remaining coals should be crushed to powder or paste before carefully replacing the din and sod. Make sure there are no soft spots in the fill-in firepit that will sink with age. Also be sure to mold the edges of well-defined chunks of sod to assure a flat surface and to give the appearance that the earth has not been disturbed. Landscape the entire cooking area by scattering leaves, twigs, or whatever originally covered the ground. It is worth the effort.

Surface Method: When there is abundant bare soil available without excavation (gopher holes, old streambeds, etc.) there should be no need to disturb the topsoil by digging a firepit. Simply spread several inches of bare soil on the ground and build a fire as usual. As with the "flat rock method," all wood should be burned completely to ashes. Crush remaining coals and scatter ashes and bare soil once they have cooled. Be careful of scorching the topsoil, and landscape the cooking area before leaving.

Firewood Selection

Select firewood from small diameter loose wood lying on the ground in order to insure complete, efficient burning.

Avoid breaking off branches, alive or dead, from standing trees. An area with discolored broken stubs and pruned trees loses much of its natural appearance.

Leave saws and axes at home because they leave unnatural and unnecessary scars and add weight to your pack.

The mark of an experienced backpacker is to use a stove when wood is not readily available or when an area could be easily damaged.

Firewood is often scarce near heavily used campsites so it should not be wasted on excessively large fires.

Scatter unused firewood- before leaving your campsite to preserve a natural appearance.

Extinguishing Fires

When preparing to leave the campsite use water and bare soil to douse the flames thoroughly. Feel the coals with your bare hands to be sure the fire is out, scatter and bury the ashes.

Human Waste

The proper disposal of human waste is important. For the benefit of those who follow, you must leave no evidence that you were there, and you must not contaminate the waters. Fortunately, nature has provided a system of very efficient biological "disposers" to decompose fallen leaves, branches, dead animals, and animal droppings in the top 6 to 8 inches of soil. If every hiker cooperates, there will be no backcountry sanitation problems. The individual "cat method," used by most experienced backpackers is recommended.

The "cat method" includes the following steps:
- Carry a light digging tool, such as an aluminum garden trowel.
- Select a screened spot at least 200 feet from the nearest water.
- Dig a hole 6 to 8 inches across. Try to remove the sod (if any) in one piece.
- Fill the hole with the loose soil, after use, and then tramp in the sod. Nature will do the rest in a few days.

When hiking on the trail, burial of human waste should be well away and out of sight of the trail, with proper considerations for drainage. The cat method is unnecessary for urination; however, urinate well away from trails and water sources. Use areas that are well-hidden, but try to avoid vegetation because the acidity of urine can affect plant growth.

If you are traveling as a group, consider a toilet pit to minimize impact.

Burning of toilet paper is preferable to burying it since it does not decompose quickly. This is essential to prevent sanitation problems from heavy visitation. If you are up to the mountain man's style, use snow, leaves, and other natural substitutes in preference to toilet paper.

Tampons must be burned in an extremely hot fire to completely decompose. When not in grizzly bear country, they can be bagged and packed out. Never bury tampons because animals will dig them up.

Disposal of Camping Wastes

Tin cans, bottles, aluminum foil, and other "unburnables" should not be taken to the backcountry because they must be packed out.

Avoid the problem of leftover food by carefully planning meals. When you do have leftovers, carry them in plastic bags or burn them completely.

Waste water (dishwater or excess cooking water) should be poured in a corner of the firepit to prevent attracting flies. If you cook on a stove, disperse water waste faraway from any body of water. Non-soluble food particles (macaroni or noodles) in dishwater should be treated like bulk leftovers. They should be either packed up and carried out or burned. Nothing should be left behind. Food scraps like egg and peanut shells and orange peels take a long time to decompose, and are eyesores to other hikers.

Fish intestines should be burned completely in a campfire. However, if there scavenger animals and birds around, and not many remains, and if the area is lightly used, then the intestines can be scattered in discreet places to decompose naturally.

Use good judgment.

Drinking Water

Better to be safe than sorry. No matter how "pure" it may look, water from streams or lakes should be considered unsafe to drink until properly treated.

The most common disease associated with drinking water is giardiasis which is caused by ingesting the microscopic cyst form of the parasite giardia lambia. Flu-like symptoms appear 5 to 14 days after exposure and may last 6 weeks or more if untreated. Other disease causing organisms may also be present in untreated surface waters. Start each trip with a days supply of water from home or other domestic source. To replenish that supply, search out the best and cleanest source, then strain the water through a clean cloth to remove any suspended particles or foreign material. The best treatment then is to bring the water to a boil for 3 to 5 minutes. Cool overnight for the next day's supply.

Another solution is to treat with iodine water purification tablets or use an EPA approved filtered water purifier device.

Bathing and Washing

Although the mountain men weren't famous for their cleanliness, today's visitors like to bathe and wash their clothes. Be aware, however, that all soap pollutes lakes and streams. If you completely soap bathe, jump into the water first, then lather on the shore well away from the water, and rinse the soap off with water carried in jugs or pots. This allows the biodegradable soap to break down and filter through soil before reaching any body of water.

Clothes can be adequately cleaned by thorough rinsing. Soap is not necessary.

Too much soap in one place makes it difficult for soil to break it down. Therefore, dispose of soapy water in several places.

Do not use soap or dispose of soapy water in tundra areas; the soil layer is too thin to act as an effective filter, and destruction of plant life usually results.

Safety and Emergency Precautions

For safety reasons, travel with a companion. Leave word at home and at your jumping-off place if a backcountry visitor register is provided. When you travel in a party, see to it that no one leaves the group without advising where they are going and for how long.

Watch out for loose or slippery rocks and logs, cliffs, steep grades, and inclined hardpacked snow fields where a misstep can cause an uncontrolled slide or fall.

Use your best judgment and never take chances.

In Case of Injury

Injury in remote areas can be the beginning of a real emergency. Stop immediately! Treat the injury if you can and make the victim comfortable. Send or signal for help. If you must go for help, leave one person with the injured. If rescue is delayed, make an emergency shelter. Don't move until help arrives unless there is more danger in remaining where you are; use extreme care in moving the injured.

Altitude Sickness

A person should spend 2 or 3 days getting acclimatized to high altitudes before hiking.

The lack of oxygen at high elevations gives some travelers altitude sickness.

Prevention: The best prevention is slow ascent with

gradual acclimatization to altitude. Beginning at an elevation of 9,000 feet, it is recommended that you do not ascend more than 1,000 vertical feet per day.

Symptoms: Cough; Lack of appetite; Nausea or vomiting; Staggering gait; and Severe headaches.

Treatment: A person with symptoms of altitude sickness should breathe deeply, rest, and eat quick-energy foods such as dried fruit or candy. Take aspirin to help the headaches; antacid pills may help other symptoms. If symptoms persist, seek lower elevations immediately. Continued exposure can make the victim too weak to travel, and may lead to serious complications.

Dehydration

Adults require 2 quarts of water daily, and up to 4 quarts for strenuous activity at high elevations. There is a 25 percent loss of stamina when an adult loses 1 1/2 quarts of water. To avoid dehydration, simply drink water as often as you feel thirsty. The "don't drink" when hiking saying is nonsense. An excellent way to determine if you are becoming dehydrated is to check your urine; dark yellow urine may indicate you are not drinking enough water.

Hypothermia

Be aware of the danger of hypothermia subnormal temperature of the body.

Lowering of internal temperature may lead to mental and physical collapse.

Hypothermia is caused by exposure to cold, and it is aggravated by wetness, wind, and exhaustion. It is the number one killer of outdoor recreationists.

Cold Kills in Two Distinct Steps

The first step is exposure and exhaustion. The moment you begin to lose heat faster than your body produces it, you are undergoing exposure. Two things happen: you voluntarily exercise to stay warm, and your body makes involuntary adjustments to preserve normal temperature in the vital organs. Both responses drain your energy reserves. The only way to stop the drain is to reduce the degree of exposure.

The second step is hypothermia. If exposure continues until your energy reserves are exhausted, cold reaches the brain, depriving you of judgment and reasoning power. You will not be aware that this is happening: You will lose control of your hands. This is hypothermia. Your internal temperature is sliding downward. Without treatment, this slide leads to stupor, collapse, and death.

Defense Against Hypothermia

Stay dry. When clothes get wet, they lose about 90 percent of their insulating value. Wool loses less heat than cotton, down, and some other synthetics.

Choose rainclothes that cover the head, neck, body, and legs, and provide protection against wind-driven rain. Polyurethane coated nylon is best. The coatings won't last forever.

Understand cold. Most hypothermia cases develop in air temperatures between 30 and 50 degrees.

Symptoms: If you or a member of your party is exposed to wind, cold, and wet, think hypothermia. Watch yourself and others for these symptoms:

Uncontrollable fits of shivering. Vague, slow, slurred speech. Memory lapses, incoherence. Immobile, fumbling hands. Frequent stumbling, lurching gait. Drowsiness (to sleep is to die). Apparent exhaustion. Inability to get up after a rest.

Treatment: The victim may deny any problem. Believe the symptoms, not the victim. Even mild symptoms demand immediate treatment.

Get the victim out of the wind and rain. Strip off all wet clothes.

If the victim is only mildly impaired, give warm drinks. Get the person into warm clothes and a warm sleeping bag. Well-wrapped, warm (not hot) rocks or canteens will help.

If victim is badly impaired, attempt to keep him/ her awake. Put the victim in a sleeping bag with another person - both stripped. If you have a double bag, put the victim between two warm people. Build a fire to warm the camp.

If You Get Lost

Someone in your party may become lost. If you or someone else becomes lost, follow these steps.

Stay calm and try to remember how you got to your present location. Look for familiar land marks, trails or streams. If you are injured, near exhaustion or its dark, stay where you are; some one may be looking for you. If you decide to go on, do it slowly.

Try to find a high point with a good view and then plan your route of travel. When you find a trail, stay on it. If you stay lost, follow a drainage downstream. In most cases it will eventually bring you to a trail or to a road. Help won't be far off.

When backpacking with children, be sure they stay with you or near camp. Discuss with them what they are to do if they become separated. They should know the international distress signals and when to use them - three smokes, three blasts on a whistle, three shouts, three flashes of light, three of anything that will attract attention.

A guaranteed method of attracting attention and getting someone to investigate during the summer months is a fire creating large volume of smoke. Green boughs on fire will create smoke. A fire should only be used as a last resort. Be sure your fire does not escape and cause a wildfire. You can be held liable for the entire cost of putting it out!

What To Do When Someone Is Overdue

Stay calm and notify the County Sheriff or Ranger in the trip area. They will take steps to alert or activate a local search and rescue organization. If the missing person returns later, be sure to advise the Sheriff or Ranger.

Credits:

Backpacking article courtesy of
the U.S. Forest Service with credit to:

Bureau of Land Management

Colorado Horsemen's Council

Colorado Open Space Council

Colorado Outward Bound School

Izaak Walton League Of America

National Leadership School

Sierra Club

Wilderness Society

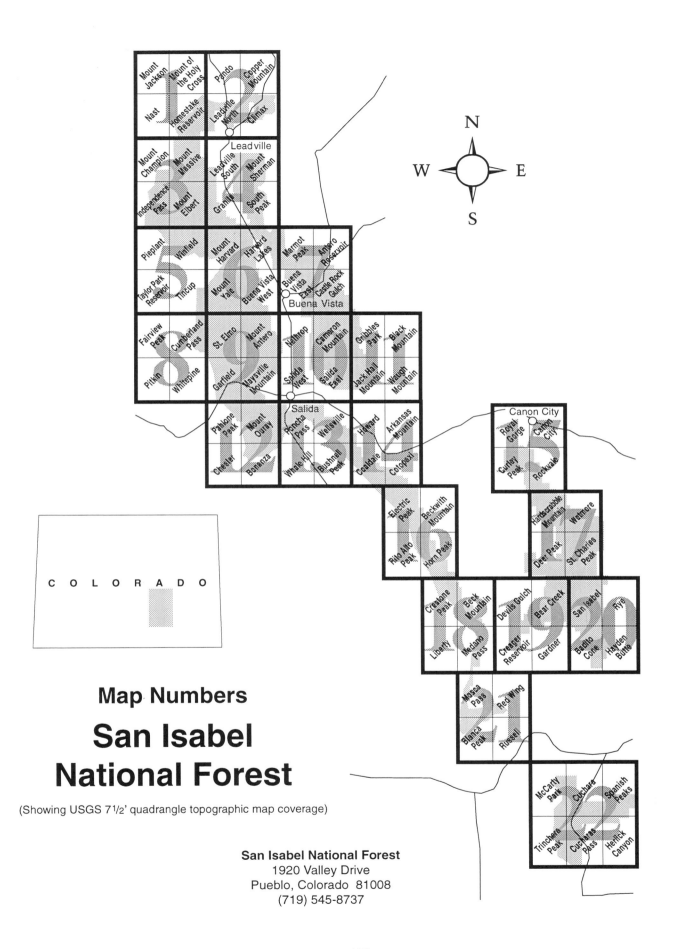

Campgrounds - San Isabel National Forest

	Site #	Name		Site #	Name
MAP 2			**Map 15**		
	1.	Father Dyer		1.	Oak Creek
	2.	Baby Doe			
	3.	Belle of Colorado	**Map 16**		
	4.	Printer Boy (Group)		1.	Alvarado
	5.	Molly Brown			
	6.	Silver Dollar	**Map 17**		
				1.	Ophir
Map 3				2.	Davenport
	1.	Halfmoon			
	2.	Elbert Creek	**Map 20**		
	3.	Twin Peaks		1.	Saint Charles
	4.	Parry Peak		2.	Southside
				3.	La Vista
Map 4					
	1.	Lakeview	**Map 22**		
	2.	White Star		1.	Bear Lake
	3.	Cabin Cove (Group)		2.	Blue Lake
	4.	Dexter		3.	Purgatoire
Map 6					
	1.	Collegiate Peaks			
	2.	Cottonwood Lake			
Map 9					
	1.	Iron City			
	2.	Cascade			
	3.	Chalk Lake			
	4.	Mount Princeton			
	5.	Bootleg			
	6.	North Fork Reservoir			
	7.	Angel of Shavano (Group)			
	8.	Garfield			
	9.	Monarch Park			
Map 12					
	1.	O'Haver Lake			
Map 14					
	1.	Hayden Creek			
	2.	Coaldale			
	3.	Lake Creek			

CAMPGROUND SITE INDEX

NAME OF SITE	MAP NUMBER	SITE NUMBER	NO. OF UNITS	ELEVATION	FACILITIES
Alvarado	16	1	47	9,000	Campgrounds, Fee, Restrooms, Drinking Water, Hiking Trail, Motorized Trail, Bicycle Trail
Angel of Shavano (Group)	9	7	20	9,200	Campgrounds, Fee, Restrooms, Drinking Water, Fishing, Hiking Trail
Baby Doe	2	2	50	9,900	Campgrounds, Restrooms, Fee, Drinking Water, Fishing, Hiking Trail
Bear Lake	22	1	14	10,500	Campgrounds, Fee, Restrooms, Drinking Water, Hiking Trail, Fishing, Horse Trail
Belle of Colorado	2	3	19	9,900	Campgrounds, Restrooms, Fee, Drinking Water, Fishing
Blue Lake	22	2	15	10,500	Campgrounds, Restrooms, Fishing, Drinking Water, Hiking Trail
Bootleg	9	5	6	8,400	Campgrounds, Restrooms, Hiking Trail
Cabin Cove (Group)	4	3	100	9,000	Campgrounds, Drinking Water, Fee, Fishing, Restrooms
Cascade	9	2	23	9,500	Campgrounds, Restrooms, Drinking Water, Fee, Fishing
Chalk Lake	9	3	21	8,700	Campgrounds, Fee, Restrooms, Drinking Water, Fishing
Coaldale	12	2	11	8,500	Campgrounds, Restrooms
Collegiate Peaks	6	1	56	9,800	Campgrounds, Fee, Restrooms, Drinking Water
Cottonwood Lake	6	2	28	9,600	Campgrounds, Fee, Restrooms, Drinking Water, Fishing
Davenport	17	2	12	8,500	Campgrounds, Fee, Restrooms, Drinking Water, Hiking Trail, Bicycle Trail
Dexter	4	4	26	9,300	Campgrounds, Restrooms, Fishing, Fee, Drinking Water
Elbert Creek	3	2	17	10,000	Campgrounds, Restrooms, Hiking Trail, Fee, Drinking Water
Father Dyer	2	1	26	9,900	Campgrounds, Fee, Restrooms, Drinking Water, Fishing, Hiking Trail
Garfield	9	8	11	10,000	Campgrounds, Fee, Restrooms, Drinking Water
Halfmoon	3	1	24	9,900	Campgrounds, Fee, Restrooms, Fishing, Motorized Trail, Drinking Water
Hayden Creek	12	1	11	8,000	Campgrounds, Fee, Restrooms, Hiking Trail, Drinking Water
Iron City	9	1	17	9,900	Campgrounds, Fee, Restrooms, Drinking Water, Hiking Trail
Lake Creek	12	3	11	8,200	Campgrounds, Fee, Restrooms, Drinking Water
Lakeview	4	1	59	9,500	Campgrounds, Fee, Restrooms, Drinking Water, Fishing, Hiking Trail
La Vista	20	3	29	8,600	Campgrounds, Fee, Restrooms, Drinking Water, Fishing, Hiking Trail
Molly Brown	2	5	49	9,900	Campgrounds, Fee, Restrooms, Drinking Water, Fishing, Hiking Trail
Monarch Park	9	9	38	10,500	Campgrounds, Fee, Restrooms, Fee, Fishing
Mount Princeton	9	4	17	8,000	Campgrounds, Fee, Drinking Water, Restrooms
North Fork Reservoir	9	6	8	11,000	Campgrounds, Restrooms, Fishing
O'Haver Lake	12	1	29	9,200	Campgrounds, Restrooms, Drinking Water, Fee, Boating, Fishing
Oak Creek	15	1	15	7,600	Campgrounds, Restrooms, Drinking Water
Ophir	17	1	31	8,900	Campgrounds, Fee, Restrooms, Drinking Water

♿ *White reversed-out symbol indicates barrier free facilities available* ♿

▲ Campgrounds	🚻 Restrooms	🐟 Fishing	🚶 Hiking Trail	🏍 Motorized Trail
$ Fee Required	🚰 Drinking Water	🚤 Boating	🚲 Bicycle Trail	🐎 Horse Trail

CAMPGROUND SITE INDEX

NAME OF SITE	MAP NUMBER	SITE NUMBER	NO. OF UNITS	ELEVATION	FACILITIES
Parry Peak	3	4	26	9,500	Campgrounds, Fee Required, Restrooms, Drinking Water
Printer Boy (Group)	2	4	125	9,900	Campgrounds, Restrooms, Drinking Water, Fee Required, Fishing, Hiking Trail
Purgatoire	22	3	23	9,800	Campgrounds, Fee Required, Restrooms, Drinking Water, Fishing, Hiking Trail
Saint Charles	20	1	15	8,800	Campgrounds, Fee Required, Restrooms, Fishing, Drinking Water
Silver Dollar	2	6	45	9,900	Campgrounds, Fee Required, Restrooms, Drinking Water, Fishing, Hiking Trail, Boating
Southside	20	2	8	8,800	Campgrounds, Fee Required, Restrooms, Drinking Water, Fishing
Twin Peaks	3	3	37	9,600	Campgrounds, Fee Required, Restrooms, Drinking Water
White Star	4	2	64	9,300	Campgrounds, Fee Required, Restrooms, Drinking Water, Fishing

White reversed-out symbol indicates barrier free facilities available

- Campgrounds
- Fee Required
- Restrooms
- Drinking Water
- Fishing
- Boating
- Hiking Trail
- Bicycle Trail
- Motorized Trail
- Horse Trail

GENERAL CAMPING INFORMATION

DEVELOPED CAMPING

Campsites are available on a first-come, first-served basis. All campgrounds limit campers to a 14-day stay in one campground. When this limit is reached, the camper is required to move to another campsite unless special written permission has been obtained. Small trailers may be used in the campgrounds. The main camping and picnicking season is from Memorial Day weekend to the Labor Day weekend. The campgrounds remain open before and after these dates but services such as water delivery and garbage collection may not be provided. Water systems are not generally operative until Memorial Day weekend when most danger of freezing is past. Post-season use of most campgrounds continues until snow depth becomes prohibitive, usually the latter part of November. Garbage collection and campground cleanup continue on a reduced-service level from Labor Day through the big game hunting seasons. Most campgrounds are heavily used in the summer months, however, Spring and Fall can offer unlimited choice of sites.

Campgrounds contain parking spurs, tables, fire grates, bulletin boards, vault-type toilets, and some campgrounds have water systems. Picnic areas contain the same facilities, with group parking instead of individual spurs. Fires may be built at campgrounds without a permit where fire rings and gratesare established. Firewood is not provided in campgrounds. Deadfall should be used for fires.Although it is usually hard to find in most campgrounds, it is easily attainable outside the campground on the Forest. Cutting of standing timber, shrubs, and other vegetation is prohibited. One note about surface water by camping sites, it may be contaminated and should always be properly treated before consumption. Pets are welcome in all campgrounds and must be kept on a leash. Horses are not allowed. Details of specific limitations, helpful information, and other regulations may be reviewed at information boards located near the entrance of each campground. Garbage collection containers are present in most campgrounds, with a "pack it in/pack it out" system in others. Visitor cooperation in keeping the campgrounds and picnic areas clean and unspoiled is appreciated. Litter and vandalism make the forest less enjoyable for everyone. Forest Service officials patrol the recreation sites and provide assistance to visitors.

DISPERSED CAMPING

Campers go hunting, hiking, 4-wheeling, and horseback riding throughout the Forest in many areas where developed camping areas are not available. Finding a site to camp and enjoy these activities in remote areas or in relative solitude has become a popular form of camping throughout the Forest. At some more popular areas, minimal toilet facilities are provided. Dispersed camping is permitted in most areas of the Forest. Some caution should be used in selecting a site because of intermingled private lands within the National Forest. Recreation maps are available at all local Forest Service offices. Personnel at each station can answer questions on specific sites. Unlike developed campgrounds which are designed and maintained to protect the vegetation, soils, and natural setting; camping in undeveloped areas require more from the camper to help keep the site in the condition in which it was found.

The following are suggestions to help maintain these areas:

Access - In areas closed to the use of vehicles off Forest roads, where developed parking sites are not provided, and where not otherwise prohibited, direct access to a suitable parking site within 300 feet of the road is permitted. Such travel must not damage the land or streams. Please select your route carefully, and do not cut live timber. Off road travel should be limited to when the ground is dry.

Human Waste - Use toilets where provided. In other areas select a suitable screened spot at least 100 feet away from open water. Dig a small hole 6 to 8 inches deep. After use, fill the hole with the loose dirt and tramp in the sod with your foot. Nature will dispose of the waste in a short time by a system of "biological disposers".

Trash - All dispersed areas are managed on a "pack-your-trash" basis. Cans, bottles, aluminum foil, and anything that will not burn should be carried out. Paper and other burnable items should be burned in your campfire. Please do not bury garbage or trash.

Water - For short trips, take a supply of drinking water from home or from another domestic source. For longer trips, boiling water for a minimum of 5 minutes is the most effective treatment for giardia cysts and other water-borne disease organisms. A longer boiling time may be required at higher elevations.

Fire - Select a site away from low limbs and clear away needles, twigs and other ground litter to mineral soil. Dig a shallow pit and line it with rocks. Keep your fire only as large as is needed to cook or heat with. Never leave the fire unattended. When you are through with your fire pit make sure the ashes are dead out then bury the pit and disperse the rocks. Whenever a stove is available, we recommend its use. This is especially true at higher elevations where it is more difficult to cook or above timberline, where wood for fuel is scarce.

CAMPGROUND FEES

PRICE OF CAMPGROUNDS (free or fee?)

FREE CAMPGROUNDS - Yes, they do exist, but drinking water is generally not provided, so bring enough for your stay. Toilets, fire rings, and often picnic tables are provided. Many of these campgrounds are "pack your trash" sites with no waste receptacles available. Be sure to have garbage bags along for your trash.

FEE CAMPGROUNDS ($$) These sites provide drinking water, refuse containers, toilets, fire grates, tables, reasonable visitor protection, and in some cases sewage disposal sites. Fees are deposited in envelopes provided at the campground entrance.

GOLDEN AGE PASSPORTS are available to any person 62 years or older. They entitle the holder and those accompanying him or her to a 50 percent discount on Forest Service campground fees. This passport is available free of charge (with proper identification) at the Regional Office in Lakewood, Colorado, and the Supervisor's Office in Delta, Colorado.

GOLDEN EAGLE PASSPORTS are not valid for National Forests but most National Park areas honor them.

CAMPGROUND HOST

In some Forest Service campgrounds you might encounter some friendly campers referred to as "campground hosts". These hosts spend part or all of their summer in a campground, free of charge. They are there to answer your questions, listen to your comments, and enjoy their stay along with other campers. If you are interested in becoming a campground host, contact the District(s) or Forest where you would like to stay.

ALPHABETICAL TRAIL LISTING

Trails are in numerical order in document.

Trail	Trail #	Page		Trail	Trail #	Page
Baker	1312	22		Mount Columbia	1494	116
Barlett	1310	20		Mount Harvard	1501	120
Browns Creek	1429	98		Mount Huron	1498	118
Cascade Creek	1303	10		North Brush Creek	1356	84
Chaparral	1301	8		North Colony	1340	66
Cisneros	1314	24		North Creek	1325	36
Colorado Trail	1776			North Fork	1309	18
A-B Tennessee Pass to Turquoise Lake		122		North Mt. Elbert	1484	114
B-C Turquoise Lake to Halfmoon Creek		124		Pine Creek	1374	86
C-D Halfmoon Creek to Twin Lakes		126		Poplar Gulch	1436	100
D-E Twin Lakes to Clear Creek		128		Ptarmigan Lake	1444	104
E-F Clear Creek to North Cottonwood Creek		130		Rainbow Trail	1336	
F-G No. Cottonwood Crk. to So. Cottonwood Crk.		132		A-B Grape Creek - Horn Creek		48
G-H So. Cottonwood Creek to Chalk Creek		134		B-C Horn Creek - Middle Taylor Creek		50
H-I Marshall Pass to Sargents Mesa		136		C-D Middle Taylor Creek - North Brush Creek		52
I-J Chalk Creek to U.S. Hwy 50		138		D-E North Brush Creek - Big Cottonwood Creek		54
J-K U.S. Hwy 50 to Marshall Pass		140		E-F Big Cottonwood Creek - Stout Creek		56
Comanche-Venable	1345	76		F-G Stout Creek - Bear Creek		58
Cottonwood	1344	74		G-H Bear Creek - Hwy 285		60
Dodgeton	1449	106		H-I Hwy 285 - Colorado Trail #1776		62
Dome Rock	1387	92		Rudloph Mountain	1327	40
Dry Creek	1343	72		Saint Charles	1326	38
Goodwin Lake	1346	78		Second Mace	1322	32
Greenhorn	1316	26		Silver Circle	1323	34
South Colony	1339	64		Snowslide	1318	28
Horn Creek	1342	70		South Brush Creek	1355	82
Indian	1300	6		South Creek	1321	30
La Plata Peak	1474	110		South Mt. Elbert	1481	112
Lake of The Clouds	1349	80		Squirrel Creek	1384	90
Lewis Creek	1331	42		Tanner	1333	46
Lilly Lake	1308	16		Tunnel Lake	1439	102
Macey	1341	68		Ute	1306	14
Mineral - Stevens	1332	44		Wagon Road Loop	1427	96
Missouri Gulch	1469	108		Wahatoya	1304	12
Missouri Mountain	1378	88		West Spanish Peak	1390	94

METHOD FOR RATING TRAIL DIFFICULTY

Four categories for degree of difficulty are as follows:

EASY:

A. Route is most level with short uphill/downhill sections.
B. Excellent to good tread surface and clearance.
C. Absence of navigational difficulties/hazards.

MODERATE:

A. Route is level to sloping with longer uphill/downhill sections.
B. Good-to-fair surface and clearance.
C. Minimal navigational difficulties/hazards.

MORE DIFFICULT:

A. Route is level to steep with sustained uphill/downhill sections.
B. Fair to poor surface and clearance.
C. Short sections involving significant navigational difficulties/hazards.

MOST DIFFICULT:

A. Route is mostly steep with sustained uphill/downhill sections.
B. Poor-to-nonexistent tread surface and clearance.
C. Longer sections involving significant navigational difficulties/hazards.

Any rating (i.e., Moderate trail difficulty) found in this guide is based on the above scale which has been established for Forest Service purposes.

ORDERING
U.S. GEOLOGICAL SURVEY
TOPOGRAPHIC MAPS

For reference Outdoor Books & Maps has divided San Isabel National Forest into 19 maps, each map represents four 7 1/2' quadrangle topographic maps (Scale: 1" = 2000 Feet). These four color maps show in detail all physical features (Water drainage, roads, towns, etc.) and elevations. The elevations are graphically shown using contour lines spaced at 20 foot intervals. The index map in the front of the guide names the quads for each map, use the index map to locate the quads you wish to order.

Ordering topographic maps from the U.S. Geologic Survey:
The maps can be purchased for $2.50 per quad at the USGS map sales located in the Denver Federal Center at 6th and Kipling, Lakewood, Colorado. (Phone 1-800-435-7627).

Ordering topographic maps from Outdoor Books & Maps:
Outdoor Books & Maps accepts mail orders for topographic maps for all of Colorado's National Forests. Using the map index within the guide locate the area you wish to visit, select the topographic map by: Map number, quadrangle name, number of topographic maps times $2.50 for each map. Add shipping costs ($2.50) and sales tax if you are a Colorado resident.

A minimum order of four topographic maps ($10.00) is required - maps will be mailed folded to 8 1/2" X 11". Please allow 15 days for delivery.

ORDER FOREST SERVICE MAPS

Forest Service maps can be purchased at any Forest Service Office for $3.00 per map. At a small scale (1" = 2 miles) a four color map shows the entire forest. All the physical features are shown except elevations, the map gives an overview of the forest. For information on locations that sell these maps, call a Forest Service office listed in the front of this guide.

GUNNISON NATIONAL FOREST TOPOGRAPHIC MAP ORDER FORM

MAP #	QUADRANGLE NAME	QUANTITY	x $2.50	AMOUNT
		TOTAL		

Colorado residents add **7.2%** sales tax: _____

Shipping & Handling costs: **$2.50**

TOTAL AMOUNT: _____

Please enclose payment to:

OUTDOOR BOOKS & MAPS
P.O. Box 417
Denver, CO 80201

Ship to: _____
Address: _____
City: _____ State: ____ ZIP: ____